PRAISE FOR *HURTS SO GOOD*

"There's possibly no one alive more qualified to write about pain than Leigh Cowart. A thoughtful, funny, and at times lyrical look at pain and its deeper human meaning."

—*Wall Street Journal*

"Informative explanations of the neurobiology of pain and pleasure, and plenty of personal reflection on the author's own relationship to masochism. Queasy readers need not apply.... Cowart's raw study offers insight."

—*Publishers Weekly*

"Briskly interweaving history, biology, and reportage...Leigh's exploded view of pain is an essential component of the excavation of pleasure for which we're long overdue.... Courageous, diverting, and written with dark good humor."

—Good Advice/Bad Gay

"Cowart has endless compassion for humans trying to find meaning and purpose while trapped in our fallible meat sacks. *Hurts So Good* is funny, explicit, and oddly wholesome."

—Caitlin Doughty, author of *Will My Cat Eat My Eyeballs*

"It's testament to Cowart's skill and charm that a book about pain should feel so joyful, that a deeply taboo subject should get such a bright and vivid airing, and that experiences that should induce winces instead trigger laughs and moments of deep profundity. *Hurts So Good* is a book of wonderful paradoxes—a rich, hilarious, and endlessly fascinating look at a world that most of us know but few of us understand."

—Ed Yong, author of *An Immense World*

"Is understanding pain—and specifically why people seek out pain—the key to understanding ourselves? Before I read *Hurts So Good*, I wouldn't have thought so, but now I'm convinced. I found myself wondering why this book didn't exist before; and the answer is, because Cowart had to be the one to do it. This is a deeply researched, blazingly written tour de force that unlocks so much of human desire, compulsion, damage, and grace. If there's such a thing as the Great American Popular Science Book, you're looking at it."

—Jess Zimmerman, author of *Women and Other Monsters*

"A thorough examination of a widely shared human experience. Cowart blends memoir with research and observation deftly, and boldly shares the gritty details of her own sensation-seeking body. Relevant to anyone seeking to understand their own relationship with physicality. A must-read for those of us who find ourselves trying to explain so many complex things about our relationships to pain."

—Stoya, writer and pornographer

"*Hurts So Good* is a high wire act during which Cowart weaves together the science of enduring pain for pleasure with their own personal, maniacally visceral experiences. The latter scenes are written so vividly—blood, guts, excrement, swollen and frozen bodies—that, at times, Cowart seems to be daring the reader not to finish. But finish you should, because there's no better exploration of masochism's appeal."

—Elon Green, author of *Last Call: A True Story of Love, Lust and Murder in Queer New York*

HURTS
SO
GOOD

HURTS
SO
GOOD

The Science and Culture of Pain on Purpose

Leigh Cowart

PUBLICAFFAIRS

New York

PublicAffairs
Hachette Book Group
1290 Avenue of the Americas, New York, NY 10104
www.publicaffairsbooks.com
@Public_Affairs

Printed in the United States of America

Originally published in hardcover and ebook by PublicAffairs in September 2021
First Trade Paperback Edition: July 2023

Published by PublicAffairs, an imprint of Perseus Books, LLC, a subsidiary of Hachette Book Group, Inc. The PublicAffairs name and logo is a trademark of the Hachette Book Group.

The publisher is not responsible for websites (or their content) that are not owned by the publisher.

Print book interior design by Amy Quinn.

Library of Congress Cataloging-in-Publication Data
Names: Cowart, Leigh, author.
Title: Hurts so good: the science and culture of pain on purpose / Leigh Cowart.
Description: First edition. | New York: PublicAffairs, [2021] | Includes bibliographical
    references and index.
Identifiers: LCCN 2021016418 | ISBN 9781541798045 (hardcover) | ISBN 9781541798021
    (ebook)
Subjects: LCSH: Pain—Psychological aspects. | Masochism. | Suffering. | Pleasure.
Classification: LCC BF515 .C69 2021 | DDC 152.1/824—dc23
LC record available at https://lccn.loc.gov/2021016418

ISBNs: 9781541798045 (hardcover), 9781541798021 (ebook),
    9781541798038 (trade paperback)

LSC-C

Printing 1, 2023

This book is dedicated to my fellow masochists, and
to C.G.C., who knew I had it in me.

CONTENTS

INTRODUCTION

"WE HAVE COMPANY COMING OVER. WE'D BETTER GET OUT TO THE SHED now."

Earlier today, I had been crying in the parking lot of a fabric store, my slide into the familiar despondency of seasonal depression proceeding apace, but at this moment I am pert and excited. I follow them barefoot on a worn path through the long, wet grass in their yard. They've been running space heaters in the outbuilding of their house in preparation for our time there: a touching gesture, but a tease nonetheless. They know how much I hate the cold, and seeing as we're standing here because I've asked them to do terrible things to me, the ominous fact that I am getting the niceties of warmth has not gone unnoticed.

I have no idea what I'm in for, other than it will hurt. I twist my feet together. They're jovial; I'm chirpy. We've had a sweet date involving poorly prepared German takeout with their mother, followed by boozy coffee drinks at a red-lit hole-in-the-wall cocktail bar down by the river. My skin is already feeling a little warm and soupy from the bowl I smoked in the kitchen. They unzip my dress. I step out of the black, low-cut scrap of fabric, and they gently remove my glasses and bra. My panties stay on because they are precisely the size of a postage stamp. It's the little things, you see.

I'm blindfolded and lying on an antique gynecological exam table, my feet corralled in the menacing chill of wrought-iron stirrups. I'm tied down to the table at the neck and under my breasts. Straining against these ropes makes me feel panicky and air hungry, so I work on my breathing exercises while they slip industrial rubber bands onto my arms and legs. My breath is already shallow, fast; I feel light-headed with anticipation. Right now, the adrenaline is from the dread, and there's a lot of it. They cultivate this feeling, a gifted curator of my experience.

They begin to snap the rubber bands. Right upper thigh near the hip. Left inner thigh near the panties. Outside of the legs, the sides of my arms. The seams of me. I start out okay, on top of the cresting waves of sensation, but soon I succumb to the reality of the pain. Early whimpering ends in a shriek, and they bind my hands with zip ties. I'm moving too much.

Now they really get going. The rubber bands bite hard; I'm seeing orange and white in the backs of my eyelids. One spot on my arm gets it bad, and every time they snap it again, I make a pitiful sound, as if hearing my voice crack when I oscillate between endurable erotic pain and actual physical agony isn't exactly what I asked for. That arm will have some purple tomorrow.

The moment fills my brain in a singular way: like if you could inflate a balloon inside my skull and make it fill the whole area, and the only thing in that balloon was just that one thing. When was the last time you were thinking and feeling one exact thing? Just one fucking thing.

Just.

One.

Fucking.

Thing.

I'm sobbing into the blindfold, and they make me come with their hands, and I'm still crying and they're snapping the rubber bands again and I'm dizzy from bucking my scrawny-ass neck against my ropes. The Hitachi clicks on, and they're making me come over and over in a way that is also agony. I am squirming away from them, going nowhere. It's too much, it's too much, it's too much.

I miss the *too much* immediately when they return to snapping rubber bands. They do this for what feels like an unknowable length of time, this back and forth: a fury of forced orgasms followed by searing pain. I presume there is a puddle on the floor, my thighs covered in thick, stinging welts. When they press their body between my legs, it is a relief; I can take more pain this way. They are like a grounding wire.

I am wet with sweat and endurance, and it's just me and them and that one thought: pain. Still snapping the wide rubber bands across my raised pink skin, they crank up the Hitachi to its highest setting, now a brutal power tool. They're shoving fingers in my cunt, fucking me hard with their hand. I am shaking on the table. Everything hurts, my body feels swollen and slippery, and they lean over me, mouth to my ear, breaking the silence with a cruel laugh.

"Is this high enough sensation for you, dear?"

I come around their hand. I feel like I am dying. Things get quiet. My body is ringing like a bell, and the crickets outside sing me back into the room.

They cut off my zip ties and release me from the table, removing my rubber bands with the tiniest of playful snaps. They take off my blindfold. They're standing over me. I look up at their big, pellucid eyes and catch their gaze just for a second before they kiss me. They pet my hair. We smile while we do this, my soggy face brushing up against their salted beard.

And just like that, I felt bad, and then better.

What do you think of when you picture a masochist?

Is it a sixty-five-year-old venture capitalist, lubed up and squeezed into a latex maid suit, his little ass cheeks quivering as Mistress brandishes her whip? Or the lithe and timid Anastasia Steele of *Fifty Shades* fame, participating in a style of coercive abuse that bears little resemblance to the world of healthy, consensual BDSM? Is it me, crying in the shed?

Maybe you see a marathoner, running an ungodly number of miles on a searing summer day, stopping to barf in the hydrangea bushes before a playground of curious toddlers, pushing ever onward toward a distant finish line. What about a hot pepper addict, cursing into their curry, pink-cheeked and their forehead wetly glazed with sweat? When I say *masochist*, do you picture someone covered in tattoos, face aglitter with metal studs and silver hoops? What about the masochists who run into icy water in the dead of winter, or slap their friends on the bottom to join a special club? (I'm looking at you, frat boys.) What about a person who bites their cuticles until they bleed? What about tire-flipping exercise clubs? What about ballerinas? Boxers? Rodeo clowns? Do you see yourself?

Do you see the pattern yet? All these people choose to feel pain on purpose. Considering the lengths to which we humans go to pursue comfort and avoid pain, why do these people do it? What do you think they are getting out of seeking pain?

The point I am making, both here and in the whole of the book before you, is that while masochism can definitely be about sex, it doesn't always have to be. In fact, very often, suffering for pleasure has very little to do with our genitals and their wily desires. And though it's true that the modern word for suffering for pleasure springs from one Austrian dude's nineteenth-century boner (see Chapter 5), the reality is that it's much, much bigger than that. Sex may be the gateway drug to getting us to talk about masochism, but masochism is so much more than kink.

Today, when I use the word *masochist*, I am describing something universal, timeless, human: the deliberate act of choosing to feel bad to then feel better. To feel pain on purpose. People have long used this tactic, consenting to suffer so that they can enjoy the deliberately engineered biochemical relief that follows painful stimuli. It's not weird. And it's not rare.

This idea of *masochism*, the human trait of feeling bad to feel better, is not an all-or-nothing proposition: in reality, it's more of a spectrum, or even a series of overlapping spectra. If ultramarathoners are masochists, what can we say about marathoners? After all, they shit themselves too, and lose toenails all the time. If participants in Polar Bear Clubs are masochistic, what about people who use the cold pools in commercial saunas? Is it masochistic to blast yourself with freezing water at the end of a shower? If dancing in pointe shoes is masochistic, what about taking pole dancing classes that leave bruises in your tender knee pits? What about LARPing—live-action role-playing—with elaborate padded weapons that hurt but do not harm?

It does not seem like a stretch to presume that all these activities, and the reasons we do them, may have something in common. After all, we are all piloting similar versions of the same haunted meat suit. There is no human experience without a body; emotions are just as physical as breathing. They are coming from inside the house, just like your thoughts and farts and various odors. And when you or I or anyone else plays with pain on purpose, we are, in a way, using the results of millions of years of evolution to achieve a kind of biohacking. To make ourselves feel better by feeling like shit first. It's fun; you might like it.

The way I see it, masochism is a very human behavior, and one that only sometimes has anything to do with sex. I won't deny that sexual masochism is one of my favorite facets of it, sure. But! As we are going to explore, masochism is, quite simply, fucking *everywhere*. For example, let's start with one of the most intense examples of masochism: the ultramarathoner. I don't think anyone here would argue that a human being who runs two hundred miles at a time without stopping to sleep isn't a masochist. One assumes that in order to undertake an endeavor as significant as a days-long marathon through the desert, the person participating in it must be getting something out of their suffering. It sure as hell isn't always money. Although some ultramarathons do offer cash prizes, money for winning has not yet become commonplace in the sport (though that may be changing). The Big Backyard Ultra, for example, a last-runner-standing race widely considered to be the most sadistic race around, has no prizes at all, merely tokens of participation. These people are pursuing this tremendous feat for the love of the experience. Through pushing their bodies to the limit, which can involve things like going dust blind and vomiting up all attempts at sustenance, on a fundamental level, an ultramarathoner pursues pain on purpose. I presume they get something out of it (you can see for yourself in Chapter 8). Why else would they do it? There must be some inner reward. It may seem paradoxical, but in this book, I'm going to show you how it works. After we're finished, you'll be able to see how all types of pain for pleasure are, in fact, very similar.

When viewed within the context of my life as a whole, my personal experience with masochism has often not been sexual. I have been a ballet dancer, an overexerciser, a serious bulimic and self-harmer, a tattoo aficionado, and a science journalist. I have used my masochistic tendencies for personal and professional gain, and I've used my masochistic tendencies to hurt myself, though these days I mostly use my masochistic streak for fun, book writing notwithstanding. What these activities and compulsions all have in common is that they involved me deliberately using my own body, the phenomena of my physiology, to feel bad, and to feel better. I choose to suffer to reap very specific kinds of rewards. Endorphins are a hell of a drug.

When I talk about masochism and suffering, I'm not talking about just any kind of suffering. A crucial tenet of masochism is that it must *always* be consensual. If it's not, it's not masochism. Period. What I talk about in this book is not generalized suffering. The act of suffering covers a much broader swath of the human experience. If one cannot opt out of suffering, then it is not masochism. It could never be masochism. A person who chooses to run through pain or lift tremendous weights until their muscles scream in agony is a masochist. A person who is forced to do these things against their will is a prisoner, or a slave. This doesn't mean that a person can't find meaning in suffering they didn't explicitly consent to; after all, many people do this, but I argue that enjoying pain outside of consensual situations is more of a coping mechanism rather than true masochism, which requires choice, consent, and autonomy. That said, many people apply the tenets of masochism to suffering in a way that hacks their bodies, to perhaps make the experience a little more bearable. Pain, after all, is intimately linked to the system in your body that provides you with such nice homebrew morphine.

The kind of masochism I am talking about is built on the idea that the masochist engaging in it can opt to end the suffering that they've chosen to inflict on themselves. An ultrarunner can stop running. A person in a chili-eating contest can refuse to eat the next pepper, though, as we will see later, the gastric consequences of their actions are nonnegotiable. A person in a BDSM scene can safeword. The ability to control the scene and stop the suffering is a hallmark of masochism and cannot be overstated. If a BDSM scene continues after a safe word is called, it's abuse. If a person is forced to keep running against their will, it is torture. Let there be no ambiguity that the suffering we cover in this book is completely consensual and under the control of the person requesting or enacting it.

Now that we are perfectly, crystalline clear on that front, I want to show you that, far from the sole purview of dungeons and bedrooms, masochism—the choice to invite pain onto our bodies—is all around us, buried in gyms and restaurants and winter beaches. It's empowering and scary, wholesome and dangerous. It's pushing your limits and feeling alive and chewing a cut on your lip until your mouth tastes like iron because doing so, in some way, seems to make you feel better. In short, masochism is everywhere. So why aren't we talking about it?

For something as ubiquitous, compelling, and varied as masochism, it's astonishing that the available literature on the topic is as sparse as it is. My research has been filled with pulpy memoirs and dry scholarly tomes. There are lurid Harlequin paperbacks. There are scholarly papers quick to judge, pearls and sphincters clutched in kind, though the tide is changing slowly. There are such kind and earnest BDSM blogs extolling the virtues of subspace and safety protocols, blogs that are truly doing the Lord's work when it comes to sharing healthy kink with the masses. But there is so little in between. Much of what has been written is from too close up: it's unable to see the broader spectrum of suffering for pleasure, too mired in jargon, too intensely solipsistic, too specific to appeal to the broadly curious.

But masochism is so much more than that. Masochism is sexy, human, reviled, worshipped, and, at times, delightfully bizarre. From ballerinas dancing on broken bones to circus performers electrifying nails in their noses, to competitive eaters horking down peppers with escalating Scoville units, masochism is a part of us. It's the people who grew up to become stunt performers, whose bruising connects them to their bodies in a way that makes them feel powerful. It's people who suffer from chronic pain and choose to find autonomy of their bodies by indulging in physical violence on purpose. It's the show *Jackass,* and it's religious flagellation. It lives inside workaholics, piercing enthusiasts, and garden-variety pain sluts. Yet the way humans suffer in pursuit of pleasure is baked into a system that shuns its discussion. Masochism has long been pathologized by stuffy psychiatrists burdened by the specific myopia of rich white dudes, but it's far more interesting than the simple masturbatory fantasies that draw us to it. The world of masochism is populated in part by some of the best and weirdest folk: people who, for a variety of reasons, do things like run en masse across deserts or into near-frozen oceans, or eat hot sauce until they long for death, or ask loving partners to beat them until they are covered with tears and snot and screaming for mercy.

At its core, masochism is about choosing pain on purpose, for a reason. And often, in my experience, that reason is to feel bad to feel better. I believe that this phenomenon—the engineering of situations in which one suffers in order to secure guaranteed relief—is worthy of a tender, hilarious, heartfelt investigation. I would know: I am an inveterate, high-sensation-seeking masochist. I am a science reporter. And I have some fucking questions.

But the first question, the quiet, insistent one, the one that started me on this journey was very simple: Why am I like this? Why do I like pain, and what am I getting out of it? It's a little, uh, weird that I get off on getting punched in the mouth by a loved one, so I thought I'd unpack that a bit. And honestly, if the answers I found were merely personal, this would be an entry for my secret journal, not a book to share with you. But that's not what I found at all. Once I started looking, I found masochism *everywhere*. Suddenly, my questions weren't about me and my various kinks and habits but about people in general. I am so thoroughly not alone in my proclivities. So, my questions got bigger.

Why do people engage in masochism? What are the benefits: social, psychological, physiological, and otherwise? What are the costs? Who does this, and why are we like this? What does masochism have to say about the human experience? Through rigorous science reporting, curious and compassionate interviews, and the lens of my own personal experiences, I'm going to take you with me on this journey to find answers to these questions. Come with me and watch the grueling end of a last-runner-standing ultramarathon, widely considered to be the Worst Ultramarathon in existence, a statement nearing absurdity in its redundancy. See the aftermath of a hot pepper–eating contest in all its beet-red agony. Watch me suffer like a howling idiot as I run into the breathtaking cold of a winter ocean, cold being my absolutely *least* favorite way to suffer. Peer inside the human brain to see how the experience of pain is created, and see how our bodies make us feel better. We'll meet circus performers, pain scientists, suspension fetishists, world-class pepper eaters, and a ballerina turned Muay Thai fighter. We'll curl up together with stories from BDSM professionals as they share how they wield torture on purpose for blissful relief. We're going to see people choosing pain, and we're going to look. Closely.

This book explores the spectrum of human masochism, the reasons we do it, and what we can learn from choosing to suffer. Maybe you dabble in it. Maybe you don't. But everyone loves to watch.

Chapter One

FROM THE TOP

SARAH LONDON TIPS HER PRETTY FACE UPWARD TOWARD MY CAMERA, flashing her spit-slick mouth guard for a photo op. It's white with pink roses; in cursive script, it says ***Bitch, please***. Her words are clunky around the soft molded plastic protecting her teeth. She tells me that in her first fight, which she lost, she was pummeled hard by a much more experienced competitor, and her mouth guard filled with blood. Moments before she tells me this, amid the intimacy of sparring bodies and the myriad aromas that come with such engagements, she had burst out laughing and squealed, "No no no!" as someone threw her down hard onto the mat. "I don't give a fuck about getting hit in the face," she tells me with a sly grin. "The body, I mind. But not the face." Pain does different things to different people.

Sarah is a Muay Thai fighter. We used to dance ballet together, but now I'm perched atop a cinder-block wall in the basement of a building in East Nashville watching her spar. Neither one of us dances anymore, not like that. It appears we have both found new ways to scratch our old itches.

I've come to see Sarah because it is impossible to talk about my own relationship with masochism without bringing up my decades in ballet. Wading into these memories feels like high-stepping through a thatch of blackberry brambles with bare legs and uncalloused soles: it hurts, though there is sweetness there too. Mostly, it just makes me want to swear a lot. When I asked Sarah if she missed ballet, she gave an immediate and resounding *no*.

The ballet world, as I experienced it, was inordinately physically and emotionally abusive, a mass of concentric circles of hush-hush hell. It was years spent cowering and starving, eternally at war with my poor, battered body, which featured breasts that resolutely refused to get smaller, no matter how small the rest of me got. It is vastly unpleasant to recall those days, though my suffering did nothing to stanch my devotion to the art form back then. If anything, it strengthened my resolve. Ballet swallowed my entire childhood and adolescence; hours of dancing after school each day gave way to leaving home for a residential conservatory program in high school. I spent whole summers at weeks-long intensives. Ballet consumed me in a way that nothing else ever has, or ever will. I loved it desperately. I know she did too.

I bring this up because I am constantly trying to answer the impossible question of why, exactly, I am the way I am. Why I like to choose pain. Specifically, did ballet make me a masochist? Or was I simply well suited to the grueling discipline of the art form because of something intrinsic to my core personality, the nebulous *you*-ness that becomes solid and nameable by kindergarten? (Two things can be true at once, and my guts tell me that the answer to both of these questions is yes.) When it comes to my complicated relationship with pain, I used to feel the kind of shame that gets yoked to a sense of uniqueness, like my burden was mine and mine alone, but I don't anymore. Now I see that hurting on purpose to feel better is no novel thing. It's everywhere, inside of hot sauce bottles, lurking in cold pools, dripping onto studio floors, and wafting through the air in the Nashville MMA gym where Sarah is trading kicks to the rib cage with her boyfriend. If this book has taught me anything, it's that I'm not special. Truly, a comfort.

Part of the not-specialness stems from the basic, shared biology of a healthy human body. For the most part, our bodies all do the same thing when it comes to processing and feeling pain; the similarities in mechanics create the universal experience of *OUCH*. It goes a little bit like this: When someone stands *en pointe*, their entire body weight pouring into the tips of their toes, the body sounds an alarm. It's the same as when a person takes a well-muscled shin to the outer thigh, as Sarah's legs are turning red in protest of: the body responds, and loudly. The nervous system fires strong and clear, with nerve cells called nociceptors rushing the alarm up to the brain, the electric current that is the message zipping along the wet noodles of our sensory apparatus. In response, the brain must take into consideration the context of the signal and creates a symphony of pain that is shaped by things like emotional state, level of surprise, and previous history of such events. From here, the body releases many signals and chemicals, including its own homebrew morphine, thanks to the endogenous morphine system. You see, *endorphin* is a portmanteau of "endogenous" and "morphine." The drugs are coming from inside the house.

It is generally understood that endorphins feel good, and they get you high. So, when I talk about feeling bad to feel better, I mean it very literally. Pain can lead to feel-good chemicals and I, like so many people, am all too willing to exploit this fact for my own benefit.

On the mats before me, Sarah takes a knee to the side of her body and winces, giggles. Otherwise, her face is a placid sculpture of concentration. She tells me that they make fun of her for laughing when she gets hurt, but I get it. We were both taught how to push through pain and how to derive pleasure from it.

Sarah and I started dancing together in middle school. We were both apple-polishing baby ballerinas with fastidiously concealed streaks of mischief. We stood in class near each other, holding the backs of our thighs apart so our legs looked skinny in the mirror, sweating and heaving through hours and hours of classwork. When we weren't mimicking pristine dancing robots, we smoked pilfered cigarettes and stayed up all night to learn Ludacris lyrics, like normal hormonal wildlings.

There is something about the cult of ballet that is hard to impart to an outsider. I can very easily describe to you how terrible it was at times: the drunk ex–New York City Ballet principal dancer who hurled a chair at me, the man who threw me out of class repeatedly because he was disgusted by my brand-new teenage breasts, the director who pretended to expel me from my prestigious boarding school because he wanted a good laugh at my tears. I can tell you about the dancers who passed out during sweltering classes in the bowels of a Manhattan summer intensive and how we were instructed to just roll them out of the way and keep going. The horrors, those are easy to share.

But the good things? The things that kept me coming back over and over again? Intangible, transcendent, addictive. And ultimately, for me, for a long time: worth it.

The next day, Sarah scoops me up from the gym where she works to take me to an all-but-abandoned mall where she trains Muay Thai. I follow her down a graffitied corridor, her extra-large, neon-green gear bag dwarfing her petite but tautly muscled body, giving the impression that I am trailing a really buff, adorable turtle. The hallway opens into the mall, which is almost entirely empty, save for a quinceañera supply store and Chonburi Muay Thai. The mall is closing entirely, forever, in two weeks, and some of the stores left trash and old merch behind in the haste of their exits. The bathrooms do not work, but I hear there is still a functional food court. I am too interested in the bodies warming up on the floor to go do any exploring. The remnants of a retail clothing shop line the walls of the gym (you know, those little metal bars that shelves hook into?), and there's a Pepsi cooler filled with bottled water and a half-empty two-liter of Coke. The mats are shaped like jigsaw pieces and fit together neatly.

"It's just like ballet," she tells me, her purple hair pulled back into a ponytail. "You learn the choreography and you get hurt doing it. It's the exact same thing." And with that, she's off to do warm-ups with her boyfriend, who is also a Muay Thai fighter. The two of them lie on their backs on the mat, swinging their legs in wide circles, loosening their hips in unison. Both of them are densely muscled, compact, which seems to be the trend among many of today's class participants. When Bruce, the instructor, comes over to introduce himself, shaking his hand sends sparks of alarm up my arm. It feels like it is entirely made of wood. I've never felt a hand like this before. It is just completely and unmistakably solid. When I tell Sarah this, she laughs. She knows.

As promised, the class really is just like ballet, with a few aesthetic differences. The clothes are tight, but there is no mirror. Bruce calls out combinations not of pliés and tendus, but of kicks and punches, and the room obeys; one person striking, one person holding the pads, then they switch. All exercises are completed on both sides of the body, for equal muscular development. The combinations escalate in intensity, starting slowly, then pushing through a sweaty corridor of pain and resolve. When the students learn new moves, there is an inevitable awkward period, as their bodies figure out how to mimic the necessary motions, coolly juxtaposed with the immediate grace of the instructor. The familiar pattern of a fumble, followed by the thrill of success. The students sharply perform their tasks but without the monastic silence of ballet class. Instead, some of them grunt and shout "ooh wee" to acknowledge a hit, their faces deep in concentration. The thud of the pads is incredibly satisfying, and I have little doubt that Sarah could break a few of my ribs with a single kick. She gets a cramp in a butt muscle and frowns, rubbing it out with a gloved hand. I laugh to myself. I watched her do this move twenty years ago in the mirrored halls of the North Carolina School of the Arts. The similarities between the art form we grew up devoted to and the activity unfolding before me are staggering. The secret language of the discipline, the ritualized movement of the group, the perfuse sweating: it all makes me intensely and immediately nostalgic for ballet class. Only this is ballet class without mirrors. This is ballet class with hitting.

I watch Sarah out there working her ass off. She's a gifted physical mimic and a tenacious athlete, looking every bit in her element as she kicks and sweats and laughs through a hard strike to the shoulder. This laugh comes with a grimace, though. She stops and rubs it, wincing, garnering a hug from her sparring partner. As she works, little glimpses of her past come through; the way she sharply keeps her weight on the balls of her feet and out of her heels, the ease with which she bends over to stretch her hamstrings and rolls up through her spine to standing, small tells that only a fellow dancer would likely notice. She gently kicks a much taller man in the head, and he smiles at her.

At the end of class, they do conditioning, which is to say they kick each other a bunch to get tougher. First, the blows land on the outer thighs, right in the meaty expanse of the rectus femoris. Then, with arms up, the kicks move to landing on the side of the body. Bruce calls them *love taps*. Faces contort around the need to remain static, as if a curled lip or clenched jaw could mitigate the force of a padded shin on a tender rib cage. The pain, the sense of accomplishment, the endorphins, the sheer fun of cultivating an expansive willpower, I get it. I get why a talented dancer would find a home in combat sports. How healing it must be to take the internalized violence of ballet and turn it outward.

It looks like so much fun.

But isn't that just why I came here? Sarah reflects back to me confirmation of my own reasoning about myself, how I got here, why I'm like this. We both danced intensely after we lost touch, and toiled mightily in the ballet realm. We both married young, with boring-to-bad first marriages that evaporated instantly once we started to become Actual People and not disconnected automatonic ballet survivors. We both spiraled out in the end days and aftermath of ballet. And we've both found ways to build back into our lives the facets of ballet that kept us in it all those years. The parallelism of our lives, revisited after twenty years of silence, is striking.

I look down to my lap and see that my shorts are riding up. Under the fluorescent light, the tops of my inner thighs are yellow and speckled with rings of mottled, purple bite marks. There are so many that the chaotic markings do not read as teeth. Even after I leave the gym, after I hug Sarah goodbye, I cannot stop thinking about how ballet shaped us both.

For me, ballet was formative in countless ways and no doubt helped deliver me to all sorts of masochistic hobbies in my adult life. And though, in my experience, dancers tend to be masochists, there are many, many other people who indulge in pain on purpose who have had no exposure to shiny satin torture shoes and shouting leotard-clad Russian women with long sticks and short tempers. But there's no questioning that those shiny torture shoes changed my life.

I got my first pair of pointe shoes when I was twelve.

It was, without question, one of the most hotly anticipated milestones of my childhood, a source of great obsession and longing, going back as far as I can remember, and probably stretching beyond that too. Whereas many children focus their excited dread onto harbingers of puberty like warbling voice boxes and secret hairs, I was more concerned about when I would get my pretty torture shoes than when my breast buds would pop. I was monomaniacal in my pursuit.

It's important not to start pointe work before the muscles are strong enough to stabilize the foot *en pointe*; pointe work too early can result in broken bones and lifelong damage. So I trained hard and often, both in class and in secret, doing relevés in my bedroom at night, in the shower, while I brushed my teeth, anywhere I could get away with the smooth, slow *up and down* of my body, heels pitching forward and back as my calves labored under strict orders. After eight years of dance classes and desperate prayer, it was finally time. My teacher gave me the blessing I'd waited for my whole tiny little life, and I ascended into clouds of pure elation.

It's critically important for any pair of pointe shoes to fit perfectly but especially the first pair. For conventional pointe shoes, the hard box (where the toes are) is made of layers of cardboard and fabric, glued together kind of like papier-mâché. The shank, the hard underside of the shoe, is a piece of stiff leather. The rest of the shoe is like a soft ballet slipper. Feet work inside the shoe to roll up through tiptoes to standing, balancing on the big toe, closely clipped toenails perpendicular to the floor. (This mechanical motion of the foot and the need to use materials that will form themselves to the shape of the foot through use is why pointe shoes break down and must be replaced when they go soft.) There is no padding in the shoe, and neither ribbons nor elastic are sewn onto the shoe before purchase. The outsides are covered in satin, and my insides quiver to think of them, one of the few deep nostalgias from my youth.

The shoe itself represented the culmination of everything I had ever wanted in my just-over-a-decade of living, so even though I knew it was going to hurt, I was excited to get fitted. After all, a proper fit helps instill proper form. If the shoe isn't perfectly snug, slipping and bagginess can de-stabilize the foot, leading to injury. If the shoe is too tight, the foot cannot be articulated, and the painful and often disfiguring art form of dancing *en pointe* is impossible. Teachers told me that padding makes it hard for danc-ers to *feel the floors*, their tone of voice implying that dancers who use it are somehow less. Some people use bits of lambswool or the more modern thin gel pads to cover the toes, but I was determined to be the kind of person who didn't need it. I used snips of medical tape and sometimes a square of single-ply toilet paper, folded around my toes like a sheet of gift wrap. As you can imagine, I was a terribly relaxing child.

Finally strong enough and with child bones ossified enough, I stood wait-ing in the dark backroom of a decades-old dancewear shop next to the bas-ket of clearance leotards from the 1980s and in front of a wall of dusty tap shoes and character heels. The stooped octogenarian motioned for me to sit, then took my feet in her hands, examining them closely for size and structure before taking some measurements and scuttling into the even deeper recesses of the store to gather boxes.

Squinting again at my feet, she asked me to stand and do some calf raises. She watched, assessing my body mechanics without comment, then complimented my teacher (not present) for her mindful assessment of my strength and readiness. Reaching into one of the boxes, she pulled out a pair of a peachy-pink Chacott Coppelia II pointe shoes and handed them to me. My blood pressure went nuclear with anticipation, hands tingling, chest flushing hot under the dim lights. *Oh my god.*

I slipped the satin fantasy objects onto my twelve-year-old feet, my toes bare except for makeshift socks cut from a pair of tights that bagged around my ankles. I stood flat-footed for her assessment. She pinched my heel, hooked her finger under the satin, and pressed the toe box before nodding toward the mirror. It was time.

I took my last invigorating steps as a pre-pointe student and stood on the mat at the practice barre. Fingers gingerly resting on the wooden support, I bent my knees and sprung to my toes, ankles solid, knees strong, fully finally on pointe. My breath caught hard in my chest, the sensation from my feet crashing into the realization that getting everything I dreamed of really would come at a cost. My desire for this moment was not enough to blunt the pain, which I'll describe as *what if you took your shoes off and kicked the wall with your big toe as hard as you could, over and over and over and over again, and then you kept doing it until your toenails turned eggplant purple and fell off?* (Spoilers!)

I nearly blacked out from the pain, but I would be damned if I'd show an ounce of regret. I pliéd and relevéd again, going up and down and up and down and *up up up up up.* I smiled big, triumphant. My feet would hurt the rest of the day, the week, the month, just years of foot pain stretched out before me, but who cared? I was a fucking ballerina. I'd been allowed to join this very niche, very beautiful pain cult. I lifted one foot to rest under my knee *en pase.*

Mathematically, being *en pointe* on just one foot, with all one's weight on the bony tips of the toes, represents a force of around 4,100 newtons per meter, which is equivalent to having the weight of an entire horse, or an entire grand piano, crushing down on a single pointed toe. I have heard from dancers and doctors alike that, to the uninitiated, the pain of standing *en pointe* is enough to make a person pass out.

And there I was, initiated and absolutely fucking giddy for it.

I took to pointe work with rapt enthusiasm, regularly bloodying my toes practicing relevés late at night in my bedroom, muting the *tap-tap-tap* of my magic shoes with carpet and sweaters so as not to draw attention to my nightly ritual. It wasn't the pain that I loved; to be honest, it really fucking hurt. I cried, a lot. But the crying, I did that alone. Ballet is supposed to be beautiful, and I wanted that beauty for myself. Every time I danced *en pointe*, I felt bad, and then I felt better.

My pointe work improved too. While my poor feet regularly turned into raw hamburger, toenails falling off, blisters bursting open and weeping, strange round calluses pushing proudly up from my softness, my ability to regulate my response to the pain got stronger. I could dance through pain, bleed through my pretty pink pointe shoes, rehearse well past dinnertime and into the dark of evening, only to wake in the night when my bedsheets ripped away fresh scabs from my oozing feet stuck to floral printed cotton-poly blend. The only shoes I could stand to wear, outside of the pointe shoes themselves, were flip-flops. It hurt too much to peel socks off, and if I was going to be in the studio from one till eight, I needed to air out my wounds as much as possible. I went to a ballet boarding school for my later years of high school, but we did our academics in the morning at the local public high school, so my non-ballet classmates were regularly treated to a full view of my bloody feet.

By the end of my career, I'd danced through missing toenails, broken bones in my feet, a torn rotator cuff, tendinitis so bad that my ligaments began to tear, head trauma from getting kicked under the chin while onstage, and a back spasm so severe that it cracked one of my lumbar vertebrae. All of it hurt. But I kept at it. It's not that the pain went away, just that my tolerance for it was bolstered by my desires and complimented by my superiors. In hindsight, I wish I had been less reckless and punitive with my body.

The freakish tolerance for the *pain in my feet* thing didn't really go away after I quit ballet. Years later, at a party, one of the hosts mentioned that she had a pair of ballet boots for photo shoots. She couldn't even stand in them, much less walk in them, but she said that I was welcome to try on the fetish item if I so desired. At this point I was in my early thirties, nearly a decade away from my once-beloved pointe shoes (for the ballet dancers and ex-ballet dancers who are reading this, my go-to shoes were Grishko 2007s, super-soft shank), but old habits die hard, and I was more than willing to try the boots on. First, I stood in them. Exquisitely painful, as promised. Then I took a lap around the room for old time's sake, feeling my toenails bruise in real time. Still got it.

But, my god, I must confess: in spite of all that, when I stub my toe, I cry.

How can this be? How is it that I could be so tough, such a belligerent taskmaster to my own body, my own feet, and yet I tolerate an unexpected table leg to the big toe with no more grace than any other random barefoot wanderer? Pain, on some level, feels so intuitive, so knowable. *Of course I know what pain is*, we say. *Pain is when something hurts.* But what is it really? Sure, we know it when we feel it. It feels obvious, immutable, a universal constant akin to death and taxes.

But it's not. Pain is an entirely subjective experience, each and every time, cooked up fresh by a brain eager for gossip of the outside world and desperate to stay safe within it. *What's going on? Have I felt that sensation before? Am I in danger? Am I hungry, sad, aroused, or tired? Angry? What do I expect to happen? What can I see, what can I smell, what can I hear? Any threats? How do I contextualize the data coming in from the nociceptors? Is it time to sound the alarm or time to chill? Is there tissue damage happening? Could there be tissue damage happening? Oh my god, am I safe? Am I going to continue being safe?? Could I be safer???* The brain asks these questions—and so, so many more— and then, using all the information it has gathered up and assessed, creates the experience of pain as it sees fit.

Pain is no simple thing. It's not a switch that gets flipped, on/off. It's more like a frog in the swamp of your conscious mind, adding its voice to the chorus of all the other sounds jockeying for position. Only sometimes, this frog can make a sound that blasts through everything else like a fucking foghorn. Like vision, taste, and hearing, pain is another easily persuaded sensory experience from the wilds of your perceptive and reactive facilities. Pain can mean danger, but pain itself doesn't necessarily mean that the body is in danger. And a lack of pain doesn't necessarily mean that the body is unharmed. Pain can be acute, chronic, diagnostically relevant, diagnostically irrelevant, enduring, incapacitating, boring, maddening, used for fun, and wielded in pursuit of some of the greatest atrocities ever committed by humankind. Pain keeps us safe. Pain ruins lives. And the more closely I examine it, the more questions I have. What do we really know about pain?

There is no way to know exactly how much someone hurts. As of yet, there is no method for quantifying how much pain a person has without asking them, no numeric score that an outsider can assign, no way for a laboratory technician to divine the painful truth through chemical reagents and whirring centrifuges. There are no tests doctors can do to examine the "pain" part of the brain. There is no single pain part of the brain.

Each individual painful experience is created based on a slew of factors and can be hard to predict. As we'll see below, the experience of pain is *always* subjective, crafted by the mind itself and subject to all kinds of outside influences, including anxiety, threat level, emotional state, previous memories, degree of anticipation, and sexual arousal. Our inner life and our surroundings not only affect the experience of pain, they inform it and are critical to the creation of the sensation.

So . . . what is pain?

———

Dr. Lorimer Moseley stands on a small stage at TEDxAdelaide for his talk, "Why Things Hurt." He seems at ease, making jokes with the crowd in his Australian accent, wearing black jeans and a slate-blue button-down. Shirt slightly unbuttoned, face slightly stubbled, head shaved. "I want to tell you a story that will explain to you the first three years of the neurobiology of pain that you would study at university."

The story begins with our Moseley walking in the bush, which he mimes across the small stage. He is a senior principal research fellow at Neuroscience Research Australia (NeuRA), and his work with chronic pain treatment is revolutionizing patient care. As he reenacts that fateful stroll through the Australian wilderness, there is a nearly imperceptible disruption of his gait, kind of like the walking version of a hiccup. It's so slight that he does it again, making sure we see it.

"Biologically, I'm going to tell you what happened just then," he says, referencing the small skip in his motion. "Something touched the outside of my left leg in the skin. That activates receptors on the end of big, fat, myelinated fast-conducting nerve fibers, and they stream straight up my leg, like *whoosh*," he says with an onomatopoeic flair. The signal enters his spinal column, then whooshes again up to the brain, where it delivers its urgent message:

"You'vejustbeentouchedontheoutsideofyourleftlegintheskin."

Moseley intones the sentence as a breathless single word, much to the delight of the crowd. The human body is covered with these intrepid messengers, and the fast ones are ON IT when it comes to being touched. If something makes contact with the meat sack, Brain simply must know about it. Safety first!

On that walk in the bush, whatever stimuli activated Moseley's fast myelinated fibers also activated his slower free nerve fibers, the nociceptors. But, on that fateful day, these messages went unheeded.

"[The message] gets to my spinal cord, and that's as far as it goes. And it says to a fresh neuron in my spinal cord, 'Ahhh, you've just been, um, something dangerous has happened on the outside of your left leg in the skin . . . mate.'" This time Moseley delivers the message in a laid-back drawl befitting the unmyelinated fibers it rode in on. And so, he continues, the spinal nociceptor takes the message all the way up to the thalamus, nonchalantly notifying it of danger on the outside of the left leg. This signal isn't traveling on myelinated fibers like the fast one was, so it's coming in more slowly.

Now it's brain time. Moseley tells the crowd that, at this point in the process, the mind must assess how dangerous the whole situation really is. To do this, "it looks at everything."

His brain asks itself, *Have we been anywhere like this environment before?* Well, yes, of course, he has been hiking in the bush before. Building on that, his brain checks for any memories of similar sensations happening on his lower leg while walking in such an environment. Has that happened before? Of course it has; such little scrapes are part and parcel of hiking in a sarong.

Dr. Will Hamilton is a psychologist who specializes in chronic pain. He explains Moseley's predicament, his gentle voice ringing clear through the phone: "Basically, the sensation goes from his leg to his spinal cord; that information gets passed up to his brain." Hamilton says that, eventually, the sensation breaks into Moseley's conscious awareness. "He knows where it is on his body. He also knows what his body is doing in space. But then his mood is really calm, and he kind of has these associational metrics: 'Well, you know, the last time that I felt that, it was just a twig that scraped my leg, and so what am I going to do about it? Well, I'm not going to pay it any attention. I'm just going to ignore it, keep going.'"

Which is exactly what Moseley does. He ignores it. "Well, your whole life growing up you used to scratch your legs on twigs," Moseley says, impersonating his own brain. "This is not dangerous." With that, Moseley gives a little kick of the leg, the organism of his body content with the completed assessment. He goes on about his trek, slips into the river for a quick dip, emerges from the water, and blacks out.

And that's how an Eastern brown snake—one of the most venomous snakes in all of Australia—almost killed a world-renowned pain researcher who barely even reacted to the bite.

The venom of an Eastern brown snake activates nerve fibers, Moseley explains, saying that at the time of the bite, his brain was getting bombarded with pain signals, "and in its wisdom, it said, 'No, no, no, no.'" Given his mood, circumstances, history, and expectation, Moseley's brain did not give him lethal-snakebite levels of pain. It gave him all the pain of a little twig scratching across the skin.

Miraculously, he survived.

Six months later, while out hiking with some friends, something touches his leg. And then he feels it. "I mean agony," he says. "I've got a white-hot poker pain screaming up my leg." He is writhing, incapacitated.

Just like last time, the fibers send a message about a sharp little sensation. "That sensation goes up to his spine; the spine says 'Yep, pass it on,'" explains Hamilton. "It gets into his conscious awareness, same thing. It's in his body map." The brain is continually cataloging sensations and threats, and we refer back to things we've already experienced to make decisions in the present, both consciously and subconsciously. So, Moseley's brain gets word of a little scratch on the outer leg. His brain assesses the situation. What does it know about what is happening right now? Ringing any bells?

Moseley's brain gets a message that something touched the outside of his leg while hiking in the bush. His brain recognizes this scenario and, having learned from last time, drops his ass like a sack of potatoes. "His pain system said, 'Well, you know, let's just be on the safe side,' and it gave him ten out of ten pain," says Hamilton. "His body collapsed. In response to a twig." There is no grievous injury of his leg this time, just a very powerful memory of last time.

"Memories and associations and meanings of the pain will really strongly dictate how much pain we actually have." Brains, our poor brains, always trying to keep us safe, despite such limited sensory inputs. Hamilton tells me that usually people assume the severity of tissue damage is equivalent to the subjective experience of pain and how intense that pain is. But, in cases like Moseley's—as well as in chronic pain patients and some traumatic injuries—that's just not always true. Turns out, there are many, many factors that influence the experience of pain, and tissue damage is only one.

Hamilton continues his explanation of pain at a steady clip, and I can hear him every so often softly tapping on his keyboard when he mentions a name, fact-checking as he goes. He's a charismatic genius who's very easy to talk to, with a warm, soothing voice that brightens with the hint of a grin on its edges. I imagine he'd be remarkably good at hypnosis as he fields my questions thoughtfully. He asks me to visualize a diagram of a person cutting their finger.

When that knife slips and the slicing happens, you wind up with a certain amount of tissue injury at the site of the incident, he explains. The brain needs to know about this, but when the signal gets to the spinal cord junction, "it basically competes with all nearby stimuli in that particular region of the body in order to decide whether it's relevant to make its way up." Up to the brain, that is.

Things like vibration, temperature, and pressure all affect the spinal cord's assessment of how important the message is. (Which is why it makes sense to rub your thumb after you slam it in the door by accident; it makes the signals from the injury compete with the signals from the rubbing, jamming the line so to speak, and perhaps lessening the pain.) "As it makes its way up to the brain, up the spinal cord, it basically has to pass through a series of different gates," gates that function as arbiters of relevancy.

The gate control theory of pain is based on the chemical mechanisms that essentially up-vote or down-vote pain signals. Hamilton describes it as "inhibitory control from higher up" that basically says *"No, that's not relevant,* or, *Tell us as much about that as possible."* But why would a brain want to ignore a signal? "Part of the evolutionary advantage to that is if you were bitten on the foot by a tiger, the survival information about that injury is not all that relevant while the tiger's still there. So it's kind of a mechanism in order to suppress the pain signal so that you can get the hell out of there, and only once it's actually then relevant to, you know, put your foot up and not use it for a couple weeks, it's only then that you're out of the danger that it actually becomes something to pay attention to." Pain gating, in its way, allows signals to be sorted, and they are sorted "up" or "down" for various reasons. Hamilton explains that there are chemical reasons, reasons at the spinal cord level, and, of course, psychological reasons.

The theory of pain gating begins to answer the question: How does the body determine when it is in danger? "Some of those gates will have to do with the body as a whole, but they'll also have to do with the person's level of threats," says Hamilton. "It's basically how threatened are you by the information? How much should you be paying it attention? Kind of think of it almost like speaker amplifiers in the spinal cord, or, you know, dampeners in the spinal cord." Also in play are any memories or associations the brain has regarding similar stimuli and what the brain knows about where the body is. Essentially, what does this sensation mean?

After passing through these so-called gates, the sensation begins breaking through into consciousness, and in comes another suite of questions and assessments. The brain asks, *What do I know about what is happening right now? What am I doing? What is my body telling me and what do I do in response?* You can think of pain as a way for the body to use a sensory experience to compel action.

"The subjective experience of pain is this mixture of things where it has to do with intensity of the sensation itself and what the person's overall activation is, how threatened they are, what their emotional state is," Hamilton says. Compounding this is the higher-level influence of memories of pain. And, of course, agency. "Do they feel like there is something they can respond to it effectively with?"

It would be hard to overstate the importance of that last sentence. When a person is in pain, whether or not they feel like they can do anything about it has an enormous effect on how their brain creates the painful experiences.

More on that in a second, but first: a quick word about terminology. When someone says that "pain is all in the mind" or "the brain creates the sensation of pain," that is true. However, the pejorative English phrase "It's all in your head" is used dismissively, as if the issue at hand does not matter because the person has simply imagined it. It is used to dismiss very real hurt and pain in many circumstances, from the doctor's office to the courtroom. So it is important that I impress upon you that when I talk about how the brain "makes" or "creates" pain, I am being extremely literal. The brain makes pain. That doesn't mean pain isn't real! Pain is extremely real. Knowing its organ of origin does not lessen the very real experience of it, no more so than dissecting cadavers denigrates the study of the body. People love the Murakami quote, "Pain is inevitable; suffering is optional," but that was from a book about leisure running written by a man who quit his job as a successful jazz club owner to become a famous novelist. I don't think the sentiment is necessarily true all the time. I do not believe that suffering is optional in many circumstances, and I certainly would never presume to tell, say, a cancer patient that their suffering was *optional*. So, when I say that pain is all in your head, it is not to imply that it is therefore not worthy of time and attention. Quite the opposite, in fact. The pain the brain makes can be all-consuming.

Speaking of all-consuming. When I ate the world's hottest pepper (spoilers!), the intensity of the experience was tempered by the knowledge that I was safe, that I chose this experience, and that it would end in forty-five minutes to an hour. Had I encountered that sensation in my mouth without the feelings of security that came from knowing what the fuck was happening, I would have absolutely thought I was dying. I felt like I was dying even though I knew I was okay! I'd had very spicy food before; I was in an area that felt safe enough—all things signifying that the threat level did not need to be even louder. The brain tries very hard to learn to recognize patterns to keep us safe. It gives us pain to teach us not to do something, to dial it back, and sometimes to avoid even nearness to the pretense of harm. That's why if you think something is going to hurt, usually it does and possibly even more than it would have had you not been so worried about it.

This has significant implications for chronic pain patients.

"Most of the time with chronic pain, what people have is central sensitivity," says Hamilton, referencing a state common in chronic pain patients where their nervous systems become trained to stay in a persistent state of high reactivity. He describes it as kind of a "better safe than sorry" approach. "Their brain tends to amplify the amount they are threatened by pain sensation," and over months and years, a kind of positive feedback loop arises. "An area will hurt; they feel threatened by it, and so they tend to avoid things that are going to cause them that same sensation. But they've also reinforced that's a really relevant and threatening thing to pay attention to," and the reinforcement can increase the pain. "Like, 'Ow, it hurts, pay it more attention; ow, it hurts, pay it more attention.'"

Curiously, part of the treatment for chronic pain is, well, pain.

"Part of chronic pain treatment falls under the umbrella of exposure and response prevention," says Hamilton. "The idea is that people are having phobic responses to pain. It's really not all that different from other forms of phobia where, because there's been a reinforced history of attempting to avoid it at all costs, it kind of gets larger and larger and larger and more threatening." In this kind of treatment, the participant slowly gets accustomed to the source of their fear. "You try to get a little bit closer to the phobic object, stay there as long as you can tolerate, and then once you've learned to tolerate being that close to it, you can get a little bit closer to it. That happens pretty naturally in a lot of exercise regimens, a lot of physical therapy, things where people are routinely reexposing themselves to tolerable amounts of pain and discomfort," he says. It happens in the bedroom, too. "I can see some parallels to masochism there in which you're kind of saying, 'You know, actually I want to increase my tolerance of pain. I want to lean into this. There are desirable reasons why I would want to be able to experience pain and not run from it.'"

Beth,* who has chronic pain from a connective tissue disorder, knows this all too well. She has fibromyalgia and places symptom onset at seven years

---

* This person's name has been anonymized to protect their privacy.

old. She also enjoys BDSM. "My life has been fringed and scaffolded by pain, sometimes becoming nothing but pain," she wrote to me. "My sexual experiences have always been a balance of pain and pleasure, because my extreme sensitivity means everything hurts but I also feel reeeeaaally good sometimes."

The indulgence in pain comes at a cost for Beth. "When I choose to be beaten nearly to bruising in a BDSM context, that's a choice to recover for at least a week. The pain is intense. It blocks out everything else like a trance state, and sometimes I dissociate. When the beating is over with, I am a sea of endorphins, merely an animal capable of experiencing sensation but not processing it or trying to figure it out. This is a rarity and a blessing worth the week of physical recovery."

Her condition is degenerative, but she works within it. "Every day I discover where the line is between progressive pain and detrimental pain, because every day it's in a different place. I guess I like pain because it's always around, and liking it is my only option if I want to stay somewhat sane."

Beth's story brings up an excellent question. With all the variables affecting the subjective experience of pain, how does masochism really play into all of this? How would these associations—the anticipation of pain and the desire for it—change how the experience of pain is created inside the brain?

"The association of someone who has associations of sexual pleasure or other forms of psychological pleasure around pain would mean that the pain is actually just less threatening," says Hamilton. Essentially, the idea is that because the participant is seeking the pain in a pleasurable situation, their experience will be more enjoyable. He also notes that pleasurable pain on purpose can easily switch into bad pain if the circumstances shift. "Obviously, masochism can have its own sort of pathology in which it can turn into abusive situations. I think there's a kind of internal switch where it stops actually being pleasurable and then all of a sudden actually becomes quite threatening." In this situation, the pain sensation changes because "the association would stop actually being one of enjoyment." This is why safe words—a specific word that, when said, causes the BDSM scene to end immediately—are nonnegotiable. Things can change fast in the moment, and consent can be revoked at any time, for any reason. If consent cannot be revoked in this manner, then it's not consensual BDSM.

Hamilton says there are very, very few behaviors that would be considered pathological in and of themselves, noting that it's much more about the effect the behavior is having on the person's life. "I can see that there are ways that masochism could be primarily a way to control negative emotional states. Part of psychological flexibility is being able to have negative emotional states and not need compulsive ways to control them. So I can just say, 'I'm angry and I'm upset today, and I don't need to do anything about it.' The masochism, in and of itself, is neutral, but then how it's functioning in the system could be pathological," depending on whether or not the behavior feels compulsive, unavoidable.

"Everyone has an ecology of emotional regulation states," says Hamilton. And for some people, pain on purpose plays a healthy, supportive role in that. "I know a lot of people in the kink community: they kind of do their own form of self-therapy in which they may have had early learning experiences that caused them to feel pretty phobic about pain, about power differentials, about safety." Pain on purpose, then, allows a safe way to explore certain types of difficult feelings. This is not to say that BDSM is inherently therapeutic or can be used to replace mental healthcare, but rather that some practitioners can exercise it as a tool for personal growth. Consensual pain in a sexual context becomes a way to "help them get a little more psychological flexibility around something that they could otherwise be rigid about earlier in their life."

Think of it this way: if the brain is the sculptor of the subjective experience of pain, masochists and other people who indulge in pain on purpose are making requests of the artist. Something that Grace* knows all too well.

"Context is everything!" she says with a laugh, her clear voice coming through the phone like wind chimes, bouncing around the inside of my parked car.

For years, Grace was a popular top in her local, vibrant BDSM scene. She tells me that the context of pain makes her think of the placebo effect,

---

* This person's name has been anonymized to protect their privacy.

explaining that people's orientation toward a painful interaction shapes their experience of it. This is called *expectation*. "That's part of the placebo response," she explains. "If I go towards this interaction thinking that it's going to benefit me, it's more likely that it will."

She recognizes a certain resonance between the placebo effect and your average BDSM ritual because of the expectations surrounding that kind of play. That sense of *Oh gosh, something important's going to happen here* is critical in shaping the overall experience of a scene. Supporting these positive expectations are the practitioner's previous experiences, as well as their desires. If it's been good before, and you think it's gonna be good now, then you are primed for enjoyment. That doesn't mean that enjoyment is guaranteed; things change in the moment, and life is full of uncertainty. It just means that you are well positioned to like it.

"There's an expectation of benefit, like, 'Whatever happens in here, I'm going to walk away from this experience, and I'm going to feel high, and there's going to be this connection that comes out of it.'" That's the fun stuff. And our ability to anticipate and enjoy such things is primed by expectation, bolstered by ritual, and strengthened by learned associations with beloved fetish objects such as latex, whips, and gags. "There's all the anticipation. There are the smells and the sights and the sounds. It's the multisensory nature of the creation of the ritual in which the pain is one thing that happens," and for the pain that is happening, context is everything, a requirement of a brain trying to make sense of it all.

Grace is quick to express displeasure at the derogatory usage of the idiom *It's all in your head*. "There's literally no way to have an experience that's not mediated by the brain." Every backache, every itchy mosquito bite, the color pink—your brain crafts everything you consciously experience.

Context is king. "As we go into these kinds of self-created rituals of BDSM, we've created this placebo, this environment and context in which we're sort of optimizing for the placebo response around the pain." Sure, it hurts. But we set it up so that it feels good, too.

And Grace would know. She's not only an experienced top-heavy switch (a scene term denoting that although she usually tops other people, she is also open to receiving in the right context) but also an experienced acupuncturist. Context, again, is everything. And with context comes ritual.

There is so much ritual in both BDSM and acupuncture, she tells me. Think of rituals as a way to signal the subconscious that something specific is about to happen. A cap and gown for graduation signals the closing of a specific education chapter. Eggnog on the shelves signals the arrival of holidays of light. Blowing out birthday candles signals hope for a new year. This kind of ritualistic association is very useful in, say, the doctor's office, where a white coat and a stethoscope signal not only authority and knowledge but also that help has arrived and healing is on the way. It is also true in acupuncture: the ritual of the experience—lights low, peaceful space, potentially strange or noxious stimuli achieved with needles—helps shape the perception of it crafted by the brain. Same with BDSM. Same with ballet.

"Placebo is about inducing the expectation of benefit from the environment," says Grace. Ultimately, it can increase the actual physical effect of the treatment itself, be it a drug or an acupuncture needle. The experience itself is just one part of how the overall thing affects you. The ritual and the expectation are the tuning knobs.

Generally, in these situations, the expectation is of positive benefit. "Like, 'Oh, if the needles hurt a little bit, I know that it's going to do me good.'" People who have the expectation that a given thing will be beneficial are often more willing to put up with some transient pain to get there, holding to the belief that it's gonna be *worth it*. Pointe shoes, BDSM, and acupuncture are all similar in that regard. They are alike in other ways, too: biologically, they work with the same mechanisms.

"When you stick an acupuncture needle in through the skin and you twist it about, you are affecting the body-mind matrix at multiple different levels," says Grace. "You're affecting it at the local level of the pain-sensing neurons and the matrix of the fascia and also the biochemistry of inflammation." Signals get sent up through the nervous system to the spine and the brain, and the body is affected at all of those levels. The needle interacts with the pain-sensing neurons, much like a thwack from a paddle would. From there, signals are passed up the spinal cord, where they are interpreted in the context of the pain matrix.

The pain matrix, she says, is concerned with three things: alarm, location, and context. It's the sensation of *ouch*, the position on the body where the *ouch* is coming from, and how you're feeling about all of that. (Like we've seen so far, a snakebite in the bush and the brush with an errant twig on a hike can do very different things to a person; context matters!) All of this intel comes together to create an individual pain experience.

But. Given the subjectivity of pain, and the importance of expectations and emotional context with regard to how we experience said pain, I had to wonder: What happens when someone wants to get acupuncture, but they are afraid of needles? Do people even do that?

"Yes. Absolutely yes. People will surprisingly seek out acupuncture even if they're a bit needle-phobic because they have read enough about it, or someone has told them about it. Or if they're in enough pain, and they've heard that acupuncture can help."

When a needle-phobic patient comes in, Grace tries to be extra reassuring, reviewing not only her history as a practitioner but also what to expect on the table. "We'll talk about the sensation of the acupuncture needle and what that will be like because that's really important from the context of whether the acupuncture has an actual neurobiological effect." She tells them that weird is good, and though things might feel transiently intense, it's going to feel mostly weird.

Importantly, she asks her patients to tell her when they feel the weird sensation, at which point she will stop moving the needle. "When you feel that weird sensation, you tell me that you feel it, and I'm going to stop manipulating the needle." It is crucial that all of her patients know that they are in control.

"Literally, the first thing I say to someone is, 'You're in charge of everything in this experience. I'm handing control of this situation over to you, and if anything happens that you don't want to be happening, you tell me, and it will stop.'" This is foundational, immovable. "Often you can really see people visibly relax just from that." It makes an incredible difference, knowing that power. She says that with certain patients, she will joke that it's just like the BDSM scene where the person has the safe word. "You can tell this experience to stop at any point."

This is particularly important in the context of people who have a history of trauma where choice was taken away from them. Directly affirming control over a potentially painful situation enables each patient or playmate to know that, fundamentally, they hold the power. "They understand that it's for their benefit, and it's within their control, and that helps them to recontextualize the experience. It helps their nervous system to relax."

I feel my throat catch. "It has to be so healing," I say, my voice echoing around the inside of my parked car. "I mean, I have medical trauma," confessing to an acquired needle phobia. When I was dying from my eating disorder (brought on, in no small part, by my years in ballet), I was an inpatient at a teaching hospital in Chicago, and on several occasions, medical students held me down for procedures I didn't consent to. At the time I was too scared to report. The one incident that stands out was the time a doctor-in-training pinned me to the bed so she could put an IV into my neck. I cried and bled all over and begged her to stop. She did not and simply made housekeeping change my ruby-red sheets afterward.

I tell Grace, voice a little raw, that acupuncture (and for that matter, BDSM) helped me to heal; opting into painful scenes helped me build resiliency and plasticity. Consenting to pain directly and on purpose has been an enormous influence in my life. It is the consent that makes the healing possible, and what I seek cannot exist without it.

In the course of my life, I have found endless fascination with the entirely subjective nature of the painful experience. I think about the physical therapists I have seen who asked before touching me and who made me feel unstoppable and strong because they taught me how to heal my body; how they could hand-wave pain away with newly focused attention and narrowly focused praise. I also think about how it feels when I need a blood draw, and the phlebotomist rolls their eyes when I explain that my arm veins are bad, how my blood pressure rockets when they don't listen to me, how hard it is to keep from crying when they blow the vein like I said they would, and how the phlebotomist is mad that I was right and the next attempt is almost always worse, such spiteful digging. I think about how that pain builds because I cannot leave. I think about how it feels at the dentist when the drugs wear off early and the doctor doesn't think it will happen that fast, so why don't you just relax.

I think about how when I was in labor and the epidural wasn't working right, a nurse came into the room and catheterized me even though I told her not to, even though I had full sensation in my pelvis and control of my bladder, even though I felt everything she did down there. I couldn't move away because my legs were completely numb from the epidural, which had been positioned incorrectly. That feeling of being trapped and hurting.

But also I think about what it feels like to be free. Hurt so many times against my will, maybe I dabble in pain for fun as a towering FUCK YOU to anyone who needs to hear it. I think about getting spanked because I asked for it, suffering contemptuously in a sauna, eating food so hot I hope it makes my nose bleed. I think about how fun it is to live in a body that can do and feel so many things, in so many different ways. How the exact same physical stimuli can cause anything from excruciation to elation to nothing at all. How mutable this one shot at life is! Each day in the meat bag is a project of turning moist electricity into buckets of feelings, and it will never stop blowing my mind.

On the phone with Grace, I think about medical trauma, acupuncture, BDSM, consent. I picture her helming a powerful kink scene, and I imagine her gently slipping needles into nervous skin. I think about how each experience of pain is singular and true and subjective, fully. I think about little baby Leigh and her shiny new pointe shoes.

Chapter Two

THE WET ELECTRICS OF PAIN

A LOT OF THINGS HAPPEN WHEN YOU CUT A TONGUE IN HALF. INSIDE the shaky video playing on my phone, gloved hands grip the tongue, fat and filled with blood. Faced with this image, it is impossible not to think of leeches, impossible not to wince, impossible not to think of *tacos de lengua* between my molars. You know the way your mouth waters when you bite through your cheek, a flood of hot pennies mingling with spit that can't decide if it's going to be lubrication for food on the way up or on the way down? Yes, it's like that.

The woman sits in the chair, quiet. Her face is serene under thin, drawn-on eyebrows, her mouth still and glittering, adorned with a Monroe piercing. The only movement is the back of her tongue, twitching like cardiac muscle or a trapped animal. Purple ink demarcates the scalpel's future path. The assistant's hands are trembling, and god, isn't that just the worst thing about this? Only one person here gets to be visibly nervous, buddy, and let me assure you, it is not you. The customer's thickly lined eyes are closed and will remain so: a marble statue in repose. The scalpel handle is plastic and green and hovers near her mouth for a nauseatingly long time, the precursor to a horror show.

Finally, we're off: It's not the blood that gets me. I see blood every month. I've birthed a baby and soaked through those enormous postpartum diapers parading as maxi pads. I've suctioned blood out of animal abdomens to clear the way for veterinary surgeons. I've spit the dredges of an errant nosebleed into my houseplants as fertilizer. What I've never done, however, is have a goddamn knife pushed through my tongue. I am questioning my decision to eat while watching this video (bland chicken salad), or possibly my decision to watch this video at all. I cannot stop thinking about the hell the brain must create to meet such an imposition.

Watching the customer willingly sit through this novel torture, I am preoccupied by the cacophony of signals that must be rushing toward the dorsal horn of her brain. I cannot tell you what she's thinking in this very moment, but I can tell you this: the nociceptors on her tongue are fucking screaming the wet electrics of pain.

Here's how that works.

The nervous system of the human body is an elaborate structure for signaling and response that, on its own, looks a lot like someone tried to construct the basic idea of a person using fractal spaghetti. Your brain, trapped in the darkness of its own safekeeping, relies on input from its myriad sensory neurons in order to react accordingly to external stimuli and keep safe the ambling meat toilet that is your sole chance at existence in this treacherous world. At the most basic level, it's the job of the peripheral nervous system—which comprises all of the nervous system equipment outside the brain and spinal cord—to tell your brain what is going on.

To do this, your body is covered in specialized cells. Called sensory receptors, these nerve cells are activated by environmental stimuli such as sound, touch, heat, light, taste, and smell. Broadly, we can categorize sensory neurons based on their respective specialties. There are chemoreceptors that detect chemicals, thermoreceptors for temperature, photoreceptors that alert to light, and mechanoreceptors that pick up on forces like pressure and stretch. However, the different types of sensory receptors themselves tend to be much more specific in the particular stimuli they respond to, and it's not always clear what the actual mechanism of action is. For example, hygroreceptors respond to changes in humidity, but researchers aren't quite sure if they do so via a mechanosensory route or if they are more like chemo-receptors, picking up changes in environmental chemistry.

I'll be honest, it's a pretty fun thought experiment to try to unpack all the different types of sensory cells that your brain relies on to construct its picture of the world around you. Chemoreceptors on your tongue report on the chemical composition of the chewed bolus lolling about your mouth. Rods and cones in your eyes describe the color and intensity of the light that you can detect. Mechanoreceptors embedded in the hair cells inside your ear shake with the vibration of sound, and your brain decodes the shaking to mean things like "screaming" or "symphony" or "ASMR paper crinkling." Baroreceptors inside your blood vessels keep an eye on blood pressure. Pro-prioceptors tell your brain how your body is oriented in space, like whether you are standing up or where your fingers are in a field sobriety test. Your ten million or so olfactory sensory neurons connect directly to the brain, each cell a specialized type of odor receptor. When molecules of petrichor or perfume or a stranger's fart waft up the nose, it stimulates the neuron to send a message to the brain, which then identifies the smell.

This all seems pretty straightforward and reasonable until you remem-ber that the nervous system is electric. As such, all of the sensory neurons must turn their observations of the world at large into an electric signal. So, for example, when that stranger's fart wafts into your nose, the smell is dissolved into your nasal mucus, where little hairlike projections from your olfactory neurons float, waiting for action. Called cilia, these projections are covered in olfactory receptors, which function kind of like a specialized lock and key system. Each cell and its cilia express only one type of olfactory receptor, though there are redundancies. When the right molecule finds the right receptor, it's the biochemical equivalent of slipping the right key into the proper lock. And inside that lock is all of the molecular machinery re-quired to turn information about an odor into literal, honest-to-god electric-ity. The electrical signal is then sent to the brain, which deciphers the input and voilà: you experience a smell.

Your charming date gently brushes your arm? Bam, electricity all the way to your brain. Do you see that beautiful sunset? It's a gift from your cen-tral nervous system, created from electrical impulses that are based on the limited spectrum of light that the human eye can decipher. Your experience of everything starts with a stimulus, which triggers the biochemistry of creating electricity, which leads to electric signals zipping up your neurons, and culminates in your brain going *Hmmmmm, what do we have here* and then, BAM, you get a sensation!

(In case you were wondering, yes, of course I am great to get stoned with. If your idea of fun involves deeply considering all the ways in which sensory apparatuses create very diverse experiences of the world, then I'm your bud. Bees see UV light! Dogs can smell through time! Butterflies taste with their feet! I can do this for hours.)

But sensory equipment is handy for more than just appreciating nice body odor or listening to trash music in secret at the gym. Sensory equipment must also tell us when bad things are happening. To this end, the brain needs to know when the body is being harmed so it can invoke the sensation of pain. It's a critical system, and one for which I dare say we should be more regularly and attentively thankful. For people born with a congenital insensitivity to pain, they face the difficult prospect of trying to navigate the world without the feedback necessary to keep them safe. How do you know you have an infection if you can't feel the pain associated with it? How do you learn your safe limits if there is no signal that they have been transgressed? People who can't feel pain will effortlessly walk on broken bones, bite through their lips, and bloody their feet in ill-fitting shoes. They often suffer severe self-inflicted injuries that dramatically shorten their life expectancy. Simply put, we need pain to keep us safe. But pain itself is anything but that simple.

It seems counterintuitive: although our experience of pain feels immediate and straightforward, it is nothing like flipping a switch. Rather, pain is a complex and subjective experience that is created by the brain itself. That's right: your brain creates the sensation of pain. But to do that, it needs to know that there's a problem. Lucky for us, one type of sensory neuron is devoted solely to detecting harm: the nociceptors.

Nociceptors are sensory neurons that sound the alarm when they are exposed to stimuli that they consider to be damaging—or potentially damaging—to the body. Structurally, they have a cell body that's attached to a tendril called an axon. This tendril snakes away from the body of the cell until it becomes a spray of free nerve endings. These tips respond to chemical, thermal, and mechanical dangers and, like other sensory neurons, turn a perceived danger into electricity. Once the baseline threshold of danger is met, say, by touching a hot stove or stubbing a toe, nociceptors sound the alarm by creating an electrical signal and sending it to where the alert can be passed on to the brain for review.

Home base for these neurons is located in one of two places, depending on which part of the body is being surveilled. For the face, the cell bodies of nociceptors are bunched up in a wad in your head called the trigeminal ganglion; for the rest of your body, these cell bodies are found in the dorsal root ganglia in the spine. The meaning of the term *ganglia* is easy to remember, because it does what it says on the label: ganglia are gangs of nerve cells, clustered together like lumpen hubs of electrical signaling.

But why are there two hubs for these cell bodies? Wouldn't it make more sense to have just one centralized location? To find out, I asked Dr. Jens Foell, at the time a neuropsychologist at Florida State University. He charitably starts with an anatomy refresher in his fast-paced German accent, reminding me that the spinal pathway and the trigeminal pathway are pretty much separate, "with the first collecting pain signals from the body and funneling them through the spine, and the other collecting information from the head [and] face." Both pathways end up in the same place in the brain; they just take different routes to get there. But why?

"In order to get parts of the trigeminal pathway into the dorsal root ganglia [where signals from the body are processed], we would have to reroute signals from the face down into the spine and then back up into the brain, which is a waste of nerve tissue." That would also mean that pain signals would have to travel farther and would take longer to be registered. "Phrased differently, our face is slapped onto one side of the brain while the rest of the body is stuck on another side of it, so these two areas take different paths into the brain."

The axons of our sensory neurons snake from these two hubs. Axons are the thin projections from the cell body that allow the neuron to play telephone with itself and receive input from the wilds of the peripheral nervous system; that is, the world outside the brain and the spinal cord. Branched out like the graceful and comprehensive roots of a plant, the system studs nociceptors all over the body. Because there are threats to the body both inside and out that we need to be aware of, these receptors are found externally (in the skin, corneas, and mucosa) and internally (in the muscles, joints, bladder, and gut). It's not an even scatter. There is a higher concentration of nociceptors on the skin than there are deep in internal surfaces, topping out at twelve hundred per square inch in your fingertips, which partially explains why a mere paper cut causes such existential anguish. It's easy to see why the body might want a better alarm system for its waterproof container than it would for the important squishy bits that, for the most part, live untouched in the dark.

Speaking of squishy bits, let's get back to that gore video. When the scalpel makes contact with the woman's tongue and begins its hot butter knife slide through the skin, it enters the receptive field formed by the peripheral ends of the nociceptors. These receptors require a certain threshold of intensity to be met before firing, and thank goodness for that. Were the danger-sensing nociceptors to misfire, the body would quickly turn into a very painful reimagining of "The Boy Who Cried Wolf," sounding the alarm vigorously when no real threat is present. Just ask anyone unfortunate enough to get into a tiff with a male platypus during breeding season. (I know it may sound sorely tempting to many of you, but seriously, don't do it: platypus venom causes a severe increase in sensitivity to pain, a condition known as hyperalgesia. It gets nociceptors to fire indiscriminately, which is a polite way of saying it makes your pain sensors scream bloody murder for no reason. The venom itself is not lethal to humans, though I imagine that some of the people unlucky enough to feel its effects might wish that it were. This leads to long-lasting pain that even morphine cannot touch. I hear it is truly hell on earth.) That nociceptors fire when there is the potential for damage is an anatomical gift; that they usually do not sing of warning when there is none to be heeded gives us a peace most likely to be taken for granted.

But, here, in the gruesome video I'm silently watching amid a bustle of cheerfully oblivious tourists in a West Asheville coffee shop, it appears that the nociceptors of our intrepid tongue-splitter are working as intended, zipping up her axons and delivering a resounding WHAT THE FUCK to her central nervous system. As the scalpel begins to slice, she sucks in a deep breath of air, her nostrils folding closed. Though the video is zoomed in very closely to what is happening in her mouth, I can see the unmistakable crinkle of skin around her eyes, signaling what I imagine to be a vastly understated wince. It's worth noting that her nociceptors would be firing similarly were she to suck on a hot poker or take a sip of bleach, but today it is the mechanical damage of the scalpel that has her pain receptors turning noxious stimuli into electrical energy. And her first wave of pain signals are heading to the brain at a zippy rate of about twenty meters per second.

That first wave gets there so fast because it is flying in on lightly myelinated Aδ fiber axons. Myelin is essentially a kind of fatty goo that helps electrical messages inside the body, be it communication between nerve cell bodies themselves or the electrical directives sent out to a muscle, get to their destination faster. The properties of myelin are often likened to plastic coating over a wire because it serves the same general purpose: electrical insulation, though the analogy is imperfect. The important thing to know is that signals traveling on myelinated fibers reach their destination faster than those traveling along their unmyelinated brethren.

Knowing that myelin increases signal transduction speed, the pain signals that come in on the lightly myelinated Aδ fiber axons are the fastest danger signals, compared to those riding in without a myelin boost. Scientists call these signals first pain. First pain is fast pain. It's big and loud and not to be missed. First pain sounds like a bazooka- and sweaty bandana–filled eighties action flick, and its effect on the brain is not dissimilar. The kinds of pain associated with messages coming in off Aδ fiber axons are those initial, extremely sharp bursts. Think stabbing, pricking, lancinating. It's that white flash when you slam your hand in a car door. First pain needs your attention, and it needs it now.

Clocking in at a tenth of that speed are messages heralding the second wave of pain signals that scientists have so poetically designated, you guessed it, second pain. These messages arrive on C fiber axons, which lack the myelination of their Aδ fiber brethren. Without myelination, second wave pain signals are conducted more slowly. The second phase of pain is longer but less intense. It signals active damage. Second pain is more pervasive; it's burning, it's throbbing, it's aching, cramping, sickening. First pain tells you that your hand is in the car door; second pain tells you that your hand is broken. First pain is triggered at a lower threshold of intensity to alert you to a problem; second pain happens when the level of intensity of stimuli is enough to recruit C fibers into the mix. First pain says a scalpel is cutting through your tongue. Second pain says that now your tongue is forked.

But where are the signals, you know, going? When the nociceptors in the tongue start to signal that damage is occurring, the messages follow the trigeminothalamic pathway, which is a richly syllabic way to say that the pain stimulus is coming from inside the face. If you were to trace this pathway, the messages would travel deeper inside your head, then dip down to the lower brain parts, but not all the way down into the spine proper. As we saw previously, in terms of the two locations for face versus body ganglia, this is different from pain signals coming from your body, which do go through the spinal cord on the spinothalamic pathway. As the names imply, both pathways head to either thalamus, one of two gray lumps residing near the middle of your head that act as a hub for relaying sensory information to the rest of the brain. I like to think of it as a kind of mailroom, where input gets sorted and sent off to the proper area of the brain for processing. Pain signals from the face and body both pass through these lil nubs (albeit different parts) and from there make it up to the somatosensory cortex at the top of your brain, where most body signals are being processed.

And here, dear reader, is where things start to get weird.

"What makes pain so interesting to study, in neuroscience terms, is that pain itself is really one hundred percent subjective," says Foell. His lilting German accent crackles through the phone, words flying out with the astonishing speed of a patter song. Soon, he'll be picking me up from the airport to go see a big magnet, but today, I'm talking to him about using fMRI (functional magnetic resonance imaging) to study pain. Pain is definitely not straightforward. "[Pain] changes around with all kinds of circumstances." Think about that for a moment. Pain is circumstantial. I think about that as I set my alarm to catch a predawn flight. I think about pain while sitting under the cruel, bright lights of a 5:30 a.m. airport terminal. I think about pain while running through the sprawling Charlotte airport to catch a connecting flight, and I think about pain while I spill hot airplane coffee on my hand. By the time Foell arrives to collect me from the time capsule that is the Tallahassee International Airport, I am ready to see how he thinks about pain.

I find that, in talking about the human body's response to pain, I am comfortable and knowledgeable when it comes to the rather straightforward process of nociceptive pain. There is a stimulus, then electric signals zip up the axons, and the information about the threat—be it cold water or twenty minutes on a spanking bench—reaches the brain. But once the brain knows about the threat, it sets about conducting and curating the experience. And that's where things get complicated. That's why I need Dr. Foell.

I find Dr. Foell waiting for me outside the Tallahassee airport. As we leave, I notice there is a severed plastic hand in the back seat of his car, still fresh in the packaging. Halloween is next week, so I don't think much of it, until later, when I find out the hand is for a bar-friendly version of the "rubber hand trick." To do it, the rubber hand is placed in a lidless box on top of a table, while the participant's real hand is placed beneath it in another box. Both hands are stroked for a bit, simultaneously, and at the same speed. After a few minutes of this, many people's brains begin to think of fake hand as self. The fun, then—the trick—occurs when you suddenly stab the fake hand with a pencil, and the hapless participant screams.

Tallahassee is hot and bright, a stark contrast to the cool, damp mountains I've left behind. Herds of medical school students move en masse, their white coats signaling an adherence to the ritual of their chosen profession. Foell buys me a coffee from a campus shop, and we sit together in the open-air courtyard, southern sun burning through the air. Feeling the cancer star sizzle on my unaccustomed skin, I sheepishly express my desire for shade. We move immediately. Foell is a consummate host, eager to answer questions and make my Florida experience as pleasant as possible. I scratch a bugbite.

I'm here to see FSU's research-grade fMRI machine to better understand how scientists study the ways in which the brain creates the experience of pain. I know what the nervous system does to alert the brain to danger, but what happens once the brain starts to respond to the threat? What, exactly, is going on up there? And how the fuck do we know this?

Specifically, I'm interested in a German study where the research team, led by Dr. Herta Flor, used big, big magnets to peer into the brains of masochists. The resulting paper, "Contextual Modulation of Pain in Masochists:

Involvement of the Parietal Operculum and Insula," published in the delight-fully named *Journal of Pain* in 2016, is dense and intimidating, and I am awed by it. Flor and her team were putting masochists in their MRI machine to look for differences in how masochist brains actually, you know, work.

Magnetic resonance imaging uses an echo-lag technique to bounce mag-netic impulses off the body. It determines what, exactly, lies below the surface of the skin by calculating the differences in how long it takes the different tissues to realign with a steady magnetic field after a magnetic disturbance. Functional magnetic resonance imaging is a special type of MRI that is used to assess brain activity. As Foell explains, "The only thing that functional MRI looks at is a difference in the oxygen rate of the blood at any part of the brain."

Foell is careful to specify exactly what this means, reiterating that the machine looks at oxygen rate, not blood flow, though the two things are related. Related, but not the same: for the machine, it doesn't matter how much blood volume is there in the brain; what matters is whether the blood is currently oxygenated or not. "The idea of our understanding of the brain is that [if] one area is very active, it's going to use up more oxygen," Foell says. Increased activity means an increased demand for oxygen means there will be more oxygen-rich blood flowing to the parts of the brain demanding it. "There's going to be a little curve while the oxygen is being used up and the blood redistributes oxygen molecules, basically." A curve of activation. "You can see, oh yeah, there's a lot of oxygen use, and then it sort of falls back to baseline. That's what happens, all the time, all over your brain, but it's stronger in areas that are uniquely activated for a specific purpose."

Imagine that you are looking at a plain old wall. Think institutional, cin-der block. If you then pick up, say, a comic book, the areas of your brain that are equipped to process more complex visual stimuli ramp up. Even though this part of the brain has been active the whole time, reading a comic book requires more work than staring at a wall, so more oxygenated blood is shut-tled over to the site of activation. It is this increase in (and use of!) the oxy-genated blood that an fMRI machine can detect, which tells us which areas of the brain are more active during specific tasks and emotional states.

Like pain.

So, where exactly is pain created and experienced inside the human brain? Well, that's a good (complicated!) question. There is something researchers refer to as the "pain matrix," as mentioned by Grace in the previous chapter. The pain matrix is not, strangely, a heavy metal band or a porn parody of a Wachowski film, but instead it describes areas of the brain that have been observed to consistently respond to nociceptive pain. This includes the anterior cingulate cortex, a collar-shaped part of the brain involved in autonomic functions such as blood pressure and heart rate, while also playing a role in jobs like attention and emotional regulation; the thalamus, which, as I mentioned before, is instrumental in sending sensory signals to the cerebral cortex for processing; and the insula, a part of the brain deep inside the fissure that separates the frontal lobe (home to our executive functions) and the temporal lobe, a region close to our ears that is important for memory and language. Like many structures in the brain, it seems the insula is involved in lots of different tasks, from love to pain to addiction to emotion, and it appears to be important for present-moment awareness. Given this, it makes sense for it to be active during painful stimuli.

But wait! The idea of the pain matrix is somewhat contested because the areas that respond to painful stimuli also respond to other kinds of input. With this understanding, it appears that perhaps the pain matrix is more responsible for general attention-grabbing needs and might not be pain specific. It doesn't mean we should discount the areas that light up with oxygenated blood on the ol' fMRI scans during scientist-sanctioned, volunteer-approved torture sessions. It just means that, well, discerning the meaning of brightly colored blobs of action is much more nuanced and difficult than merely identifying and naming landmarks.

This isn't the first time researchers have butted heads over where pain lives inside the brain, in part because we simply do not yet have clear answers or straightforward techniques to definitively assess what the actual fuck is happening when the body is in pain. Or when the brain creates the experience of pain, as it happens. Don't get me wrong: there is a lot we do know! The world of pain research is teeming with researchers working hard to piece together the emotional and physical puzzle of both acute and chronic pain responses. It's just that brains are complicated.

Think about all the different things the brain has to do when the body is in pain. First, it has to direct a physical response: *Do I move away? Stay still? Run? Scream?* It also has to assess how to emotionally respond to the pain. Was it expected? Does responding emotionally to the pain change the odds of survival, like inside a war zone or after a shark bite? Should I stop and sob or spring to action? Threat-level assessment during the pain response is crucial to survival, as is staying calm enough to resolve the situation. Things like love, stress, anxiety—they all change our perception of pain. Hell, just being able to see the pain inflicted upon your own body helps lessen your perception of it. (Careful readers will note the utility of a blindfold for increased pain sensations during a BDSM scene.) So, it seems natural that there wouldn't be a single focal point of activity in the brain during the experience of pain, given the multifaceted, coordinated response the brain must churn out every time we stub a toe.

Which brings me to that German paper that looked at activity in the brains of sexual masochists. Given our understanding of the pain matrix, it's important to remember that different areas of the brain are involved in the various aspects of pain. As Foell told me, "While a solid handful of areas [in the brain] are related to pain processing, some of them seem to respond to different elements of pain." Given the different areas involved in processing and creating the sensation of pain, it follows that perhaps the brain of a sexual masochist might light up differently from that of a person who does not enjoy pain in their pleasures. Asking this question was the premise of Flor and her team's research: Do sexual masochists process pain differently from how other people do? And if there are differences, are the differences in how the masochists' brains process the physical pain signals themselves, or in how their brains respond emotionally to the signals?

"The somatosensory cortex, for example, clearly responds to the physical pain stimulus, so its activation wouldn't necessarily be expected to change along with the emotional value of the pain," Foell said. In plain language: "That area shouldn't respond differently between masochists and nonmasochists, because the physical pain they experience is the same; it's just the emotional evaluation that should differ between those groups."

So, if the researchers had found differences between masochists and nonmasochists in the somatosensory cortex, one interpretation could have been that masochists process pain in a way that is fundamentally different from how nonmasochists do. But that's not what they found. In terms of this expected, fundamental processing, the two groups were alike. Instead, the researchers found that the masochists had more activity in their superior frontal gyrus and medial frontal gyrus, areas that are involved in memory and cognition. "It seems that basic sensory processing is the same, but the cognitive understanding of the pain differs." That is, the pain feels the same, but the masochist's emotional response to the pain is different. This tracks with what Grace said about expectation and ritual being important components of the overall experience. "This result is pretty much what would be expected if masochists have prior masochistic experiences that they recall during the experiment, and that makes the situation feel familiar to them." Broadly, having pleasant connotations with pain and sex lights up the brain in a slightly different way than if the association isn't there. In the study, everyone received pain while looking at various images, some BDSM-related. And wouldn't you know it, the brains of the masochists looked a little different during this test.

As Foell explained to me, the masochistic and nonmasochistic brains process pain stimuli very similarly, and the apparent differences seem related to prior experience and familiarity with pain. That is, from the vantage of the fMRI machine, a lot of the basic processing looks the same. It's just that there seems to be an additional activation in the brains of the masochists. And that activation could have to do with having a desire for pain that comes with being familiar with suffering for fun. Bit of a chicken/egg—that a masochist brain might light up more because it's more familiar with the material—but it's an interesting finding to build on, nonetheless. Something different is happening; we're just not sure why.

There are some caveats to this study, of course. The sample size was small, only thirty-two subjects, sixteen self-identified sexual masochists and sixteen people who were not. I also note that the criteria to qualify as a sexual masochist seemed awfully stringent. It read as follows: "To be included in the study, they had to consider themselves to be a masochist with a clear preference for the submissive role and more than 50 percent of their overall sexuality had to be acted out with pain-related masochistic activities (e.g., flogging or whipping) in real life, not exclusively online." This participation criteria would exclude me, the author of a book on masochism, on the basis that I am a switch who enjoys both giving and receiving pain, but also never mind the fact that if I included pain play in over half of my "overall sexuality" (whatever that means), I would be hospitalized in days. It's also important to consider just how difficult it is to study people with sexual desires that are considered taboo or fringe-y. Most masochists, if they even qualify for the study, probably aren't super stoked to look at porn inside an fMRI machine in front of some scientists, so in that regard, the study participant pool self-selects for some level of sexual exhibitionism as well.

As Foell and I were finishing up in the basement at FSU, the fMRI machine still chugging through its phantom scan, the conversation turned, as it does, to masochism. Without missing a beat, a colleague of his looked up at me from her work and smiled. She told me that she swears hot sauce helps with her migraines.

Linking brain activation and nociception seems straightforward, and in many ways it is, but it is not without complications. Complications that Foell is happy to delve into. When he gets excited, his speech becomes dappled with little forte accents, a helpful guide to parsing the dense information we're discussing. And that information is the difference between how the body signals pain and the experience of being in pain. "That's why people distinguish between pain and nociception, because nociception you can measure. You can measure nerve activation, you can extract the whole nerve and stimulate it and quantify the signals and so forth." This is simply not so with pain. "Just because there's nociception in your body does not mean there's pain. You could be in a war zone, and you could be severely hurt, but you're not feeling it." Alternatively, a person could be highly anxious about receiving pain, which would sensitize them to the experience. That is, their brain would create a more intensely painful experience, based on things that have little to do with nociception.

"Nociception and pain: sure, they belong together somehow," he muses. "But, yeah, they are different." One describes the electrical alarm system of the body; the other describes whatever response is dispatched to attend it. His voice rises with excitement. "And you can feel pain without nociception!"

Excuse me, what?

Foell's PhD focused on phantom limb pain, he tells me. "The limb might be gone, there might be no nerve signals whatsoever, and people still perceive pain." This is very common and can happen very explicitly. "They say, 'Oh, my little finger hurts!' when the whole arm is gone and has been gone for twenty years. That means that, somehow, pain happens in the brain and nociception happens someplace else, which makes it weird and confusing."

And that's not the only weird thing about pain. Our other senses, when exposed to constant stimuli, graciously dim to the input, a kindness at metal shows and perfume counters alike. Called adaptation, it's a way for our bodies to parse signal from noise. As far as the ol' brain is concerned, if the sensory input isn't changing, it becomes less important because we roughly know what is going on. Take vision, for example. Our eyeballs jiggle constantly as a way to keep refreshing the image we see so that we can keep seeing it. Frogs, on the other hand, do not have jiggly eyes and therefore can only see motion. To test this yourself, you can gently hold your eyeball still with your fingers (through your lids, please!). If you are able to get a good, soft grip on your eye, cover the other eye, hold your head still, and the picture in front of your eyes will fade away because there is no new input. This is an excellent party trick that will have a group of adults looking like complete weirdos in no time.

Our sense of pain appears to be the only sense that does the opposite.

"If you're in a situation where somebody will give you some slight pain all day, you would think that your body would adapt to that, and you would feel less pain, but the opposite is true. Your nociceptor cells will become more sensitive to the pain." This immediately calls to mind my long hours of ballet spent in bloody pointe shoes, feeling my hamburger meat feet get pulpier as the day progressed. "If you want to put an evolutionary theory on top of it, you can say that makes sense because your body is telling you to get out of that situation, so it raises alarm after alarm."

Moreover, sensitization to pain can make us jumpy and expectant when placed in similar situations. "At some point, you're going to be overly sensitive, and you might twitch away, even if [you're in] just kind of a similar situation to where you're usually being hurt." The body learns when to anticipate pain. If you are in a situation where you have been hurt before, you are likely to experience more pain because your brain knows that it is coming and remembers the threat. That's the opposite of what all of our other senses are doing, the adaptive response described above. But the brain is creating pain. And it's a creation that only gets louder the more you ignore it.

B-b-but, what about the nociceptors? What of the relentless wet electricity of the human body, carefully delivering messages and sounding alarms? How is it that so much of what we experience as pain has nothing to do with the nociceptors, yet bodies rely on them to get the whole suffering experience started? Unlike the relatively objective nature of our other senses (though, convincing arguments have been made that our senses are less objective than we'd like to admit), the translation of nociception into the experience of pain is highly subjective. Nociceptive pain pathways are nice to talk about because we understand them reasonably well; we can trigger them, observe them, and measure them. They are relatively straightforward and mechanical in a way that can be nicely summarized. I can watch a tongue being cut in half, and I can neatly visualize signals marching their orders. I can talk about which sensory nerves are firing and how quickly those messages are getting to their destination. What I cannot do is tell how the pain feels to the person experiencing it. If nociception is not an indicator of pain, what, then, is affecting our perception of pain?

"Anything that's sort of . . ." Foell's voice trails off, signaling the messiness inherent in my question. "Physical or physiological arousal or an emotional state of mind would influence pain perception. Sexual arousal, being in a war zone, being in love, just being emotionally invested in something, all of that would raise the pain threshold." As an emotional, high-sensation-seeking masochist, I am immediately covered in painful gooseflesh at the recognition. (It's delightful.) The wet electrics of pain are incredible, but they are only the beginning. Pain is biological, but it is also personal, intimate, mutable. Foell tells me that if you give an emotionally invested or sexually aroused person an objective painful stimulus "like the electric shock or the same intensity of a heat laser or something like that," they will perceive less pain and exhibit a higher pain threshold than a person who is not in such a state. Their brain will show them different types of suffering based on how they feel.

I love Dr. Herta Flor's research because it illustrates just how messy and complicated pain is in the brain. The reason I struggled so much with her paper itself, the reason I came to Florida to pick through Foell's brain and let a lab tech use transcranial magnetic stimulation to make my arm twitch, the reason I've spent hours and hours wandering through neuropsychology papers: the reason is precisely because the brain is a mess! Three pounds of soft, fatty goo running this whole show on electricity. Somehow, we masochists can hijack the straightforward, no-nonsense signaling of nociception, process it in our Jell-O noggins, and orchestrate countless flavors of pain and reward that are more than just *ouch*.

In this way, I have come to think of my experiences with masochism as a kind of biohacking: a way to use the electrochemistry of my body in a deliberate way for the purpose of curating a specific experience. Something about my response to pain is different, be it inborn or learned (or both, I suspect). It's something that allows me to craft a little pocket of joy for myself, an engineered release, be it through running a few miles uphill, getting a tattoo, or getting slapped in the face for fun until I cry. The truly fascinating thing to me is that the nuts and bolts of the pain response are so similar: in general, the meat of us works more or less the same way across the board. But our emotional response to pain? The way we as individuals experience pain? That is something that varies wildly, from person to person, moment to moment. Consent, mood, desire, anxiety, memory, expectation, love—all of these things affect the end result of nociception, the wet electrics of pain. So, many people engage in the ritual of deliberately feeling bad to feel better; once I started looking for the pattern, I saw it everywhere.

I think about the subjectivity of pain in the context of masochism, and my mind reels. How does wanting to suffer change the brain's interpretation of the act itself? Do I feel less pain because I asked for it, or do I feel more pain because I have been sensitized to getting the shit kicked out of me for giggles? The body wants to protect itself, and the cavalcade of processes that happen after injury are many: a flood of endorphins, a burst of adrenaline. Something better to follow something horrible.

Toward the end of the tongue-splitting video, the woman's face has become a pale moon spewing crimson. Gloved fingers struggle to wrangle her twitching tongue, gauze darkening in the new split of her. Hands in her mouth, she is already trying to speak. Spit and hot blood pours from her face into what appears to be a trash can. She opens her raw wound before taking a sip of water from a red plastic solo cup. She swishes the liquid around her mouth and looks up for a moment, wiggling like a puppy while her cheeks bulge with water. And then, after everything, she smiles.

Chapter Three

THE FUFFERING OF THE FAINTS

THE LILAC EDGES OF DAWN BEGIN TO POOL ATOP THE TREE LINE, SKIT-tering in on little rat feet. The smell of a musty straw mattress greets you with the day, same as your fleabites, same as the stale mouth taste of teeth wiped with a dirty wool cloth before bed. You have to shit, but you wait. The chill of the morning air is too bracing to face with such immediate and winking vulnerability. Definitely need to do it before the day heats up, though. The dry pit you've been shitting in ripens dramatically in the sunlight.

The year is 1349. You are thin and rashy, undernourished, unvaccinated. Your shoulders ache in a complete and thorough way; the toil of recent days has left a screaming missive inside the muscles that cradle the rough interior of your skeleton. A neighbor screams, unearthing a great, shrill seam of grief. You expect to soon hear the familiar clatter of a wooden wheelbarrow. You gently probe the side of your neck, fingers walking underneath the edge of your jawbone, nervously exploring. Nothing is swollen, but such risings feel inevitable. It's possible that your groin is sore because you have been poking it so much, rubbing your creases in your sleep, touching them, examining them in every sliver of privacy. The air stinks of death, the town is filthy with it. Black flies crawl in and out of noses, and no one is getting last rites anymore because who is left to bestow such a kindness?

The plague is here. Unfortunately, so are you.

It's hard to overstate just how quickly things can change. For months, the town has toiled in quiet dread, going about the choreography of daily life amid whispered messages of death that march through the German countryside. The days held promise before it got here, and life had been enriched with the possibilities of a world with a future, despite what the cautious whispers from out of town foretold. Some say that it was hard to keep going while annihilation crept around the doorstep. But not you. You found it easier to soak yourself in the balm of denial and continue the machinations of what life has always been for you: chopping wood, fetching water, making soup, making babies, losing sleep, starting fires, complaining, getting drunk, normal human shit. Before you were a gravedigger, you spent your days tending an ungrateful flock of wool sheep, using your calloused hands to skirt and scour fleece. Now you dig great ditches to fill with your neighbors, their pale faces looking up from the dirt, buboes the size of baby fists that fuse together to form a topography of loss. Mary and her little one are across the pit from each other, each silvery blue in the early stages of death. Reunion would require someone to climb down into the pit, move some lost angels around. The layer of bodies on the bottom is older, bloated, squishier. Ripe. Mary's baby will have to spend their eternal rest near their mother and that will have to be enough. You can't remember who died first, mother or infant. You'll probably bury more of her family later this week.

Are you numb to it yet? Some people take five days to die, some three, some go on the same day. How many days do you have to wait to find out how long it will take you to die? Johannes says it's a mercy to go fast, but you are unconvinced that rushing into death is the best way to meet it. Is not the last sliver of a life worth seeing? (No, not this sliver, not like this.) The hedonists drinking and fucking their way into hell, facing the end times in an alcoholic haze. You must admit, they have a point. But you hear that this is God's vengeance. Is it not our duty to repent, to save the rest?

From your position on the frayed edge of the town, you hear a low rhythmic noise, a welcome respite from the sounds of corpses off-gassing and shifting as bacterial anarchy spreads through the leftovers of their lives. The sound draws closer, and you realize that you are listening to a litany of chanting and scuffling feet. Many feet.

Like a black snake cutting esses through a grassy pasture come dozens of people, hundreds even. Perhaps three hundred! It's mostly men and they march first; women in the rear. Shoulder to shoulder, two by two, they march in unison to the church. You make a sharp path toward the sound of the church bells, which clatter urgently, announcing the arrival of salvation. You realize now what this is! You've heard about them, these dirty wanderers saving the souls of the damned.

The pilgrims solemnly gather around the stone edifice of the town's only church. They begin to form a circle, stripping themselves to the waist as they enter formation. They march around and around in a circle until someone gives a signal. Immediately, they all drop to the ground. Most people lie on their backs, prone, open. Others curl into their sides; some hold their arms akimbo in the air. One, who you later understand was representing a sinner who had perjured, raises three fingers. You are suddenly very aware of the crowds pressed together to see the spectacle. Where are the flowers you usually hold for protection? How exposed you feel without them, exactly like it feels when you dream about it.

The leader of the procession is a tall, lean spire of a man. He begins to walk through the sea of fetid bodies, posed carefully over the strange shapes they inhabit atop the cobblestones of the town square. It is the capital *M* Master, an overseer, the one who makes the groveling possible for all who want to grovel. Think of him as something of a facilitator. A facilitator with a fucking whip.

The Master walks through a sea of bodies and whips them prostrate at his feet. The church bells have stilled to silence, and the air is filled with the sound of leather striking skin. Some of his strokes leave only pink, puffy pillows of violence in their wake, welts that swell up from within like the buboes they purport to ward away. Other lashings strike with such speed that they split the skin like a waterlogged tomato, a fine mist of blood sputtering into the air, mingling with the smells of sweat and skin folds and bodies. The faces of the flagellants contort into wrinkled visages of desperation, or they relax into the slack-jawed smoothness of a trance, or they cry fat, salty tears, or they scream, or they laugh or weep softly. The onslaught of sounds and smells fills all the crevices of the town like dawn after night. It is all-encompassing.

The Master thrashes the penitents one at a time, and each waits in holy stillness for their bid to ward off the disease that is sweeping through Europe. Years later, epidemiologists will estimate that the death toll was as high as 60 percent of the continent's population, and that, at the very least, one out of three Europeans died from infection due to bacterium *Yersinia pestis*. All you know now is that no one is spared the "loathsome, cadaverous stink from within" as Welsh poet Ieuan Gethin describes it. "Woe is me the shilling of the armpit. It is seething, terrible . . . it is the form of an apple, like the head on an onion, a small boil that spares no one."

Neither the healthy, nor the young, nor the pious can escape the Pestilence. It is said that the Great Mortality is a punishment from God, and with it the people are dying alone: you live in a culture steeped in the importance of last rites; dying suddenly, without absolution, is a fresh horror. The desperation that builds when there is no answer, no recourse, no mercy is a choking fear, and people are gripped by the need to do *something, anything*. Which is why the Brethren of the Cross are here. They are an answer to what will in time be known as the Black Plague.

After the Master finishes whipping the penitents, they rise to their feet to begin the frenzy of whipping themselves. This is the part that some of you've maybe heard about. The sheer mayhem of hundreds of people flailing themselves makes the air smell like that of an abattoir. Every person has their own scourge consisting of three or four leather straps, each with a sharp metal stud on the end. The flagellants seek to save humanity by offering their own suffering to God, striping their flesh in earnest. They offer, to us, to you, to themselves, the state of feeling bad as a pathway to the state of feeling better, a better life or a happier ending. The flagellants represent catharsis, hope. A way out. They are also, even by the standards of the time, absolutely disgusting. Penitents are forbidden to bathe during their thirty-three-day pilgrimages or even to change their soiled clothes.

The flagellants continue their whipping frenzy as a few members of their order sing to encourage them. According to Henry of Herford, a fourteenth-century Dominican friar, some of the flagellants drive the metal spikes of their scourges so deeply into their bodies that it takes more than one tug to pull the shards of metal out of their flesh. Many of the townspeople join the singing; others hold out cloths to catch the blood splashing off of the mass of bodies, pressing the newly ensanguined rags to their eyes. The Master walks among the faithful, beseeching them to pray for the souls of all sinners. Weak and welted, the flagellants throw themselves back to the ground, feebly utter prayers, and rise again to begin the process anew. Pain as penitence, from the Latin *poena*, meaning "penalty," or "punishment."

"The mediaeval equivalent of Airbnb!" Seán Martin exclaims over email. He's a prolific writer and author of *The Black Death*. The book features a detailed account of flagellant movements during the eponymous plague: if anyone has logistical questions about Germanic flagellants in the late fourteenth century, Martin is your guy. He taught me that the flagellants solved the problem of lodging by crashing on proverbial couches across Europe. "Travel in the Middle Ages was always something of a hardship in itself, as travelers would have to seek lodging either in inns or in people's homes." Imagine this lot showing up on your doorstep, hands clutching whips and crucifixes, eyes feverish with religious zeal and probably also actual fevers.

But why was this happening? Why were people migrating across the continent and whipping themselves in the streets?

"Flagellant movements during the Middle Ages sprang up in response to crises, such as the Black Death in the 1340s, or the first recorded appearance of a group of flagellants in Italy, in 1259," he explains, mentioning that the early Italian group formed in response to famine. He tells me that, in both cases, the groups sought to appease the Almighty with a show of piety. "Mortification of the flesh—in various ways—has a long history in the Judeo-Christian tradition (hermits in caves and deserts, the Stylites, who lived on top of pillars, monks wearing hair shirts, fasting, etc.), so this would have been fairly standard in terms of what a mediaeval person might do to try and avert disaster." It's appropriate that Martin uses the term *mortification*. From the Latin *mortificāre*, meaning "to put to death," the term first began circulating in English in the late 1300s as the plague flagellants were crisscrossing Europe. The concept of mortification of the flesh, rejecting the body and earthly desires for a more spiritual path, is found throughout the Old and New Testaments. Saint Paul says in Corinthians, "I keep under my body, and bring it into subjection." The symbology of Jesus's torments suffused nearly every part of religious life at this time, so it's little wonder that some believers would take up the whip along with the crucifix. That is, pain on purpose was a common part of life in those days.

"The extent to which people thought things were the result of sin is quite alien to modern sensibilities, but seems to have been fairly widespread," Martin writes. "If things went wrong, you had to do something about it to please God—that seems to have been a fairly basic mindset." He notes that there were more common forms of penitence for medieval Catholics, like prayer and confession. Not everyone went in for the more dramatic displays of repentance. "The flagellants were at the extreme end of this spectrum."

The flagellants weren't alone in their extremity; they were just one of the myriad penitential flavors in Christian Europe. And it wasn't just whipping, either. Penitents were doing all kinds of masochistic shit in the name of the Father. Take, for example, female mystics of the 1300s to 1500s. Temporally overlapping with the flagellants, these mystics (who were mostly women, but not always) punished themselves with starvation and gagging on sticks. Their suffering was "basically wanting to make yourself small and sort of godlike in that you have no need," says writer and relic expert Elizabeth Harper. "It's a rejection of your body."

The image of a nun wasting away recalls *sokushinbutsu*, Buddhist monks who fatally self-mummify through asceticism. Or the Jain practice of *sallekhana*, a voluntary vow to slowly starve to death. Mortification of the corporeal body in pursuit of spiritual ascendance appears throughout history and all over the world.

Whereas the suffering of the flagellates was public and external, the suffering of functionally anorexic nuns like Catherine of Siena was private, internal, forging a direct link to God rather than making a dramatic appeal before the masses. This link had a practical purpose: Women traditionally had no power in the Catholic church, but piety itself was valuable at the time. And with such value comes power. Harper cites this bodily rejection as key to the rise of mysticism. "The idea is that you could gain spiritual power and currency in a patriarchal system by just cutting everyone out and saying your relationship with God is direct." Harper says that the mystics appropriated the theology that said their flesh is inherently sinful and gained power through a "you can't hate me more than I hate myself" devotion to penance. "It came out in lots of ways that seem sort of crazy now but that at the time were necessary, I think, to shock the higher-ups into believing." Believing in the purported direct link to God, that is.

I ask if flagellation and starvation were the only types of religious pain on purpose from this period. "There's definitely other types of penance," says Harper, citing for example the practice of wearing a hair shirt, or cilice. A cilice is like the diabolical counterpart to your favorite set of cozy PJs. It's a homely garment made of coarse, prickly animal hair meant to scratch the skin and leave the penitent rather, well, chafed and uncomfortable. Loads of famous Christians wore them, from Saint Patrick to Charlemagne, and the practice continues today for many during Lent. But does making yourself itchy rise to the label of pain on purpose? To ancient Jewish people, the answer is no. Wearing a cilice during periods of mourning or penitence was a brilliant workaround to their proscription in the Torah against self-harm. Leviticus 19:28 makes the rule pretty clear: "Ye shall not make any cuttings in your flesh for the dead, nor print any marks upon you: I am the LORD." And as we all know, the LORD of the Old Testament was not fucking around. Similar statements come up in the Talmud as well. People don't tend to forbid activities in their holy books for no reason. Worshippers of Baal and other ancient deities predating the Abrahamic faiths were said to ritualistically cut their own flesh and tattoo themselves with ash pressed into open wounds. So the Jewish ban on self-harm probably had more to do with preventing heretical practices than protecting its worshippers.

At this point in the interview, I'm imagining things aren't going to get much more surprising, but boy am I wrong. "Purposely eating disgusting things is a big thing for a lot of saints," Harper says, noting that their religious masochism was not limited to hunger alone. "Saint Catherine of Siena sort of famously ingested the pus from—"

The line goes dead. My struggling mountain cellular reception takes a big shit, and I am left hanging, desperate to know what kind of pus this nun was eating. I get Harper back on the line. "The last thing I heard was,

'Catherine of Siena famously . . . ,'" I said, not wanting to seem too eager about the pus thing. (I was very eager about the pus thing.)

"Oh, Saint Catherine of Siena drinking the pus from another nun's breast tumor." *Oh.* (Good luck ever unknowing that little nugget!)

Catherine also famously starved herself to death, saying that she could exist solely on the eucharist. She even stopped drinking water. "She basically killed herself." She was one of the first mystics that other women painfully modeled themselves after, to mixed results. "What's interesting is that some women modeled themselves after her and became particularly holy or saints themselves, and some women modeled themselves after her and wound up getting severely rebuked for doing it just a little bit wrong." This comes down to the two ways to venerate saints: *imitata* and *admiranda*. "*Admiranda* is to admire from afar and understand you're not spiritually ready to do the things that they did, and *imitata* is to actually model your life after the saint." Harper notes that some spiritual directors discourage *imitata*, but things are not as straightforward as *Do not do what the dead saints did because you'll die.* To understand why, it helps to understand the physiology of pain and suffering from hunger.

Starving, like other forms of pain, can make you feel high as a motherfucker. Harper, who like myself has a history of disordered eating, studies these mystics as a way of understanding what she's been through. We both know that though the practice of starving is ultimately detrimental to a living body, the cocktail of drugs your body unleashes in the face of real, urgent, painful hunger is a heady concoction indeed. It's dangerous.

For example, let's take a look at some research that explored an interesting similarity between starvation and, of all things, MDMA, also known as ecstasy. (I promise you, I am not making this up.) Published in 2007 in *PNAS* (*Proceedings of the National Academy of Sciences of the United States of America* if you want to be specific, but *PNAS* is more fun to say), Valérie Compan and her team at Centre National de la Recherche Scientifique used mice to demonstrate that starving and ecstasy can cause appetite suppression in the same way. Here's the breakdown:

Anyone who has taken it knows that ecstasy is a potent appetite suppressant. I once tried to eat an oily dolma out of a can when I was starting to come down and was so revolted by the sensation of it lolling about in my mouth that I almost chuckied all over the floor. Compan and her fellow researchers wanted to know why ecstasy made this happen to me and so many other who tried to eat too soon after rolling. To figure it out, she looked at an area of the brain associated with feelings of reward, the nucleus accumbens, which is chockablock with serotonin receptors. Ecstasy acts on many things in the body, including these receptors. Now, when you give that lovey-dovey club drug to mice that have been genetically modified to not have these specific receptors, the mice keep their appetite, even though they took MDMA. The drugs don't stimulate the thing causing the "don't eat" feeling, so the mice stay nibbling. And in healthy mice, when scientists stimulate these "don't eat" receptors, guess what: mice aren't hungry.

Now here's where it gets interesting. Stimulating these "don't eat" receptors also led to the release of a peptide called CART. Increased levels of CART in the bloodstream have been found in test subjects who have taken psychostimulant drugs like MDMA *and* in anorexics. And elevated levels of CART have been shown to cause animals to eat less. Basically, what the researchers found is a biological similarity between starvation (as in anorexia) and getting high on drugs (like MDMA). It's not a full report on the similarities, and there is lots of room to explore how the euphoria derived from each might be alike (or different!), but it's curious to note the finding in the context of hungry, religious nuns.

"A thing that I think about a lot is that if you're not eating, [you get] a crazy sense of energy and accomplishment," says Harper. "When I read about all these tech bros who are doing these fasts and weird diets that are basically managed anorexia, I'm just like, 'Oh yeah, of course you feel amazing,' it feels amazing," she says, referencing the intermittent fasting trend adhered to by the very thin CEO of Twitter, Jack Dorsey, much like the starving saints of yore. That is, it feels amazing until it doesn't. Eating disorders have the highest mortality rate of any mental illness.

Does anorexia feel like rolling? No, obviously not. But the fact that they share a common signaling pathway in the brain's reward center is notable and indicates a potential new therapeutic target for drug development, which is great. Anorexia is notoriously difficult to treat. It adds an intriguing note to conversations surrounding religious fasting and the euphoria that many fasters report experiencing. Especially because fasting hurts.

There seem to be several mechanisms at play in the creation of hunger. On the gross mechanical level, you have stomach cramping and strong gastric acids acting on the stomach lining. Neurobiologically, you have a complex network of neurons centered inside the hypothalamus of the brain that work tirelessly to receive inputs and send out signals about hunger. It also looks like both the endocannabinoid and endorphin systems are involved in modulating hunger and satiety signaling, as those familiar with the exogenous versions of both can attest.

Something that I think gets lost in the discussion of modern eating disorders is how very painful they are. It's easy to focus on the aesthetics of emaciation and point fingers at the fashion industry and ballet masters, but thinking of this kind of suffering only in terms of its outcome misses something foundational to the pathology. People who've had an eating disorder know it; fourteenth-century mystics knew it; the Buddhist monks who starved to the point of self-mummification knew it; people without enough to eat know it: hunger really fucking hurts. And with pain can come a lot of different feelings on the side. "There is something about pain in all of this, especially with a Catholic background, that touches on a feeling of moral superiority," says Harper.

"Being 'sure of it' is very hard to find in sort of real, everyday life," she continues. *Sure of it*, in this case, meaning feeling "sure of one's faith," but it could also have to do with being sure of one's self or one's path. Certainty is a feeling, but pain can be something of a shortcut. "When you're causing yourself pain, it lights up that part of your brain, and there is some kind of reward that's given." It's a direct link to the raw feed. It's mastery over self. It gets you high as shit.

Which is why many spiritual directors, especially modern ones, are careful to delineate between *admiranda* and *imitata*. "You can't just go getting high off not eating or flagellation yourself because that's actually a form of pride."

There it is. Corporeal experiences of spirituality, be it flagellation or starvation or piercing with skewers, make you high, sure. But DIY salvation circumvents the power and authority of the church—and their bank accounts. It cuts out the middleman, and the church hates that.

Whether going through official channels or not, all the various styles of penitents—the flagellants or starving nuns or sinners donning hair shirts—have one thing in common: guilt. Oh, guilt. I can hear the ghostly voice of my Catholic grandmother talking about it even now. Guilt: the backbone of so many faiths. But what's the connection between guilt and pain? Why do so many penitents seek out pain even when official doctrine frowns on it?

In a 2011 study published in *Psychological Science* titled "Cleansing the Soul by Hurting the Flesh: The Guilt-Reducing Effect of Pain," Dr. Brock Bastian and his colleagues examined the relationship between pain and atonement. And wouldn't you know, they found something very interesting indeed.

Participants in the study were asked to write about one of two things: an instance of rejecting or excluding another person, or an innocuous interaction. After they finished writing, they filled out a survey about how guilty they felt. Then, the fun part: they had to stick their hand in ice water. Well, some of them anyway. Control group got room temp, the bastards.

A note about this: Have you ever tried it? The ice-water thing? It's horrible! The unrelenting cold creeping in, the escalating sense of doom and wrongness, the cacophony of alarm bells urging action. Study participants who held their hand in for more than three minutes were disqualified for being able to do it for too long, and three minutes is not a very long time. This makes me immediately flash back to one particular summer in the ballet program at the North Carolina School of the Arts during which sadistic physical therapists made me sit in ice baths up to my chest to help treat a back injury. I was fourteen, miserable, teeth clattering away like a bag of bolts in a dryer, ice up to my titties in a stand-up metal bathtub, silently cursing the small rubber ducky that bobbed serenely atop the frigid slurry, mocking my suffering with its idyllic connotations of bubble baths and self-care and warmth. My friends, it fucking hurts! I do not like the cold, a distaste that will become even more apparent later in this book.

But back to the guilt study. The researchers found that the people who wrote about their guilty memory held their hands in the ice water longer, rated the ice water as more painful than the others did, and afterward *experienced a significant reduction in guilt*. Read that again. The guilty people took more pain, said it hurt more, and felt less guilty after. But why? The authors reference D. B. Morris's book, *The Culture of Pain*, which holds that "pain has traditionally been understood as purely physical in nature, but it is more accurate to describe it as the intersection of body, mind, and culture." (As the author of a book about the societal and biological implications of pain on purpose, I am inclined to agree with this assessment.) This model of thought holds that people give meaning to pain, and Bastian argues that people are socialized from birth to accept pain within a judicial model of punishment. Add in the influence of a watchful deity, and suddenly pain as atonement makes more sense.

"As we articulate in that paper, I think that more it's a relationship between pain and justice. Enduring pain can feel like it provides a sense of justice, and a form of self-punishment," Bastian says, noting that the embodiment of the punishment can be linked to penitence by varying degrees. "It's not that people explicitly say to themselves, 'I'm punishing myself with pain,' but rather [they are] going for a hard jog or doing something that's exertive and fulfills that need to restore justice through punishment." As Bastian states in the paper on guilt and pain, "History is replete with examples of ritualized or self-inflicted pain aimed at achieving purification."

There is a delightful book on the subject, prosaically titled *The history of the flagellants, or the advantages of discipline; being a paraphrase and commentary on the Historia flagallantium of the Abbé Boileau, . . . By somebody who is not doctor of the Sorbonne*, by Jean Louis de Lolme. It's a thick tome of self-inflicted pain, accounts—sometimes lurid, sometimes tedious, often moralized—of whippings and weapons and disciplines from the time of ancient Syria and Greece to the great outpouring of marching flagellants during the Black Death. But I gotta say, one of the things about it that is really going to stick with me is the usage of the letter *f* in place of the letter *s*.

Which is how *the suffering of the saints* becomes *the fuffering of the faints*, a phrase that instantaneously embedded itself in my lizard brain with the fury of a starving tick. But I digress. There are just so, so many examples of masochistic religious penitence.

There are tales of Spartan boys being whipped for an entire day as a part of their coming-of-age rituals. A painful trial, not unlike the Maasai circumcision ritual, the bullet ant initiation ritual of the Sateré-Mawé tribe in the Amazon, and the sometimes-fatal alcohol-soaked spanking rituals of some American fraternities. Once a year in Sparta, the boys "are whipped for a whole day, often to death, before the altar of Diana the Orthian, and they suffer it with cheerfulness, and even joy; nay, they strive with each other for victory; and he who bears up the longest time, and has been able to endure the greatest number of stripes, carries the day." It was called The Contest.

Keep looking, and you'll find Egyptian flagellation in festivals of Isis that Herodotus wrote about. "After preparing themselves by fasting (he says) they begin to offer Sacrifices, and they mutually beat each other." There are accounts of several thousand men and women beating each other in the street in celebration. Please take a moment to imagine three thousand people—roughly the capacity of the Philadelphia Metropolitan Opera House—swarming the sidewalks, enacting a sacred Fight Club right there in front of the gods and kids and dogs. (Would you join in? Depends on how you feel about mosh pits, I suppose.)

There's a dramatic account from Apuleius, who, in his work *Metamorphosis*, or *The Golden Ass*, tells us of Syrian priests who not only whipped themselves but also cut open their arms. "In fine, they dissect their own arms with two-edged knives, which they use constantly to carry about them." These Syrian priests hollered about other people's sins and beat the shit out of themselves, which I imagine was a riveting spectacle. Some contemporaries said that this was just a pretense, a way to raise admiration in the minds of the weak and superstitious; the priests were accused of putting on these painful shows not to appear pious but to take people's money. (These same claims were later made against the medieval Catholic flagellants, too.) There might be some truth to such claims, but who's to say where the hard line is between piety and profit? I will also note that painful spectacle in exchange for money works! The masochistic sideshow act is alive and well today, but we'll get to that later in the book.

Ooh, and let's not forget Lupercalia! This feisty ancient Roman festival in celebration of fertility and its mascot, Pan, Lupercalia was filled with nudity and lashings. Naked men roamed the streets with floggers, whipping the hands and bellies of women—some of whom were pregnant!—so that these women might find themselves more fertile or more capable of an easy delivery. Also, for those playing along at home, you should know that Pope Gelasius replaced Lupercalia in the fifth century with a Christian festival to celebrate a venerated martyr. That martyr, once canonized, became Saint

Valentine. Once the trappings of Christianity fell away from Valentine's Day, it became, for modern folks with such proclivities, a suitable holiday for floggers and whips once again.

The list feels endless, and rightfully so. Clowning flagellation by the philosopher Peregrinus, who performed his ablations in a jocular way! And there's the delightfully named Superanus, an ancient philosopher who was known to whip himself in the public baths! Nuns who whipped themselves before the election of a new prioress to brighten their understanding of things!

And it's not just history: there are many painful rituals practiced today, like Easter Sunday crucifixions in the Philippines, the procession of the flagellates in Spain, and the visually stunning Tamil Hindu *kavadi* ritual. In the latter, worshippers form a procession, carrying various forms of *kavadi*, which means "burden." Though some carry mild burdens such a jugs of milk, others pierce their flesh with long, metal skewers. Mortification of the flesh in this procession can involve pushing *vel*, skewers, through parts of the face, like the cheeks, lips, and tongue. Some even use small spears. The procession can take days.

But wait, something doesn't add up here. Looking back at these examples of religious pain on purpose, it's clearly not just about penitence and guilt, though obviously it can be. What I'm seeing here, and what I've seen throughout my research on the subject, is that it is not merely pain and penitence that are so conjoined but rather pain and *ritual* that are so deeply linked. Guilt and absolution, then, are merely a subset of a larger tradition. Why might such an intimate connection exist?

A piece of this particular puzzle lies in another paper from Brock Bastian and team. Published in 2019 with collaborator Sean Murphy, the paper, titled "Emotionally Extreme Experiences Are More Meaningful" found that rather than assigning meaning to an experience based on its valence (that is, how good or bad we perceive it to be), we instead give meaning to extreme events that inspire emotional intensity and contemplation. For anyone who has fasted and meditated for days on end or summited a challenging mountain peak, I imagine these assertions might feel self-evident, but it's always good to try to back up claims with research.

Bastian tells me that "people find emotionally intense experiences the most meaningful," going on to cite the work of Harvey Whitehouse on the ways people bond through ritual. "One thing we know from all of that work is that pain is often used in rituals in general because it signifies something meaningful. It leads people to connect around something important," he says. "It's an identity marker in many ways."

At their very core, rituals are about identity and creating meaning. We humans use them constantly, be it selling our children's facial bones to the Fae (also known as the Tooth Fairy), using Grandma's haunted china for big family meals during special holidays, decorating with bright lights during the winter, or just taking a quiet moment of gratitude before eating. There are little rituals we do every day without thinking: a skin-care routine, a certain drive to work, the way we start our days. There are rituals that land-scape the wilds of our lives: weddings and handfastings; birthday parties and blessingways; funerals, bachelor parties, divorce orgies, all of it. What, then, is the link between pain and meaningful, self-defining experiences?

There is something special about pain, and I will let you in on this rather obvious secret: pain lives, undeniably, in the present moment. Bastian calls it "a shortcut to mindfulness." In *The Body in Pain*, Elaine Scarry puts forth the idea that pain can tear down our identity, obfuscating our sense of self. Roy Baumeister, who has a significant body of work on the topic, refers to masochism similarly, as a way to escape the self.

What this sounds like to me is transcendence.

I think about my time in the shed. *(The moment fills my brain in a singular way: like if you could inflate a balloon inside my skull and make it fill the whole area, and the only thing in that balloon was just that one fucking thing. When was the last time you were thinking and feeling one exact thing? Just one fucking thing.)*

"Where our sense of self is very much in the past and the present and the future," Bastian says, "pain brings us immediately and brutally into the present." Pain puts us in a place of mindfulness. It is a way to be here now, to test our endurance, to earn something. "The notion of spiritual transcendence is often situated in the moment, and I think pain is a brutal force that brings us into that moment," he says. "Some people find a sort of spiritual experience attached to pain."

Bastian says that he believes there's a "very clear underlying neurochemical hedonic pathway through which pain can release and increase a sense of pleasure," being careful to note that there are also more abstract ones as well.

It is a delight to hear an expert softly muse about the corporeal experience of masochism, of the things that a deliberate painful experience can create. "A sense of really establishing justice, a sense of creating meaning in the world, a sense of social connection. All of these things are positive and can create pleasure, but in the moment, in that kind of immediate effect is also that kind of painful process."

I believe, through research and interviews and personal experience, that using pain for its own sake is an everyday part of being human. I think the capacity to seek out and benefit from pain is built into us, embedded in the looping chemical user manuals that come installed in our rented primate bodies.

Bastian writes in his book *The Other Side of Happiness*, "This intuition that enjoying pain is not only abnormal but also immoral is strong. Yet, as with many of our intuitions, when you stop and really analyze the assumption, things become a little less clear. We get pleasure from many things in life, and seek out activities that provide this for us, so what is so different from getting pleasure from pain?" The two are so inextricably linked and overlapping that, over the years, I've stopped trying to delineate cleanly between the two. "Understanding the link between the experience of pleasure and the experience of pain, however, reveals that *all* of us get pleasure from pain."

And honey, truly, everybody does it. "While we might see a sinister element in those who engage in more extreme or sexualized masochistic practices, there is a sliding scale including a great many commonly accepted forms of pain enjoyment."

We look to the extremes—the flagellates, the mystics, the monks—and we see reflected back to us our own capacity to suffer on purpose, our whispered proclivities for *just the right kind of pain*. We see pain reducing guilt, pain creating meaning, pain signifying a transition, and pain marking a special occasion. In this kind of pain on purpose, there is, at once, the sacred and the profane. Which is how a chapter that starts with the desperate flagellations of religious zealots finds its way to the concrete floor of a mostly empty arena, where high above a woman dangles from her knees.

The Asheville Tattoo Convention is sparsely attended, especially for ten o'clock on a Saturday night, especially considering that there are performers hustling onstage. Around me, people idly poke at their phones while artists slip needles into their skin, each in such a ubiquitous manner of recline that the one person who brought pillows and blankets and books to their rib cage session looks like an antiquated genius. All of these people, paying money to endure something painful, to get something in return. Permanent art, endorphins, autonomy, the reasons are many, but they all have one. And there's no denying that tattoos hurt. The familiar whirring of tattoo guns makes my skin itch for another, but I ignore it. Besides, I'm working and not in the mood to sprawl out on a sticky massage table in the basement of the Harrah's Cherokee Center.

This building has a strange structure that kind of looks like a brutalist concrete flying saucer. The outside was once used as a movie set; inexplicably, they added palm trees and used it as a stand-in for a Mexican airport. I once performed "Spanish Chocolate" in the *Nutcracker* upstairs in the expansive, warmly lit auditorium with the Moscow Ballet (and Sarah, no less!). There is a hot tub sale going on somewhere down here, too, which I know about because someone put up dozens of yellow signs to advertise this fact every foot or so along the sidewalk outside, but I see no hot tubs. I don't know what else to tell you, the energy is just weird in here.

Onstage are a handful of performers in expensive costumes, who are probably not getting paid enough to take them off for this tepid crowd and who certainly aren't getting anywhere near the compensation due to them for having to endure the nightmare of cold fluorescent lighting. A sideshow performer is packing up her gear along the side of the stage; underneath her fishnets are several tiny white squares of toilet paper, dotted in blood, from where audience members staple-gunned small bills to her body. A tall, glowering woman dressed as Wednesday Addams gives a large, chuckling man in shorts a lap dance, but not before binding him to the chair with electrical cables and covering his face. Another woman eats a whole banana onstage. The finale involves an enormous balloon. Like I said, the energy is kind of weird. And I'm here to see people fly.

As soon as the dancers are finished, Steve Truitt and his human suspension team begin to move some of the few remaining audience members away from where they will be hanging people by hooks, creating a clear swath of bare concrete floor through the sparse "crowd." Stevie Nicks's "Edge of Seventeen" blasts through the arena, tinny and echoing around the empty room, largely ignored. I wonder what it's going to be like to watch something like this in person. For such a roadkill-scraping, surgical-video-watching, high-sensation-seeking person, I am actually a very sensitive, soft pudding baby, so I am aware that I might have a bad reaction to watching someone hang from their knees. I mean, shit, I get the watery-legs, scared-of-heights tears when I watch cartoons involving hot-air balloons. Masochism takes all forms.

I feel a low-level thrum of adrenaline as I watch the riggers manipulate the ropes and talk seriously to each other with stooped heads and flatly lined faces. I don't know this yet, but the flyers are going to be high enough from the ground that a fall to the floor would be catastrophic. The preparation for this feat is of deadly importance. And I have a front-row seat. Well, me and the second grader next to me. Being here feels strangely like church.

The first person to fly has two piercings in his upper back, meaning he'll hang in what is called a *suicide suspension*, a reference to the upright position of a hanged man. Instead of hanging from literal hooks, he has two wide bar piercings over his scapulae. The riggers attach him to the ropes, and he walks onto the open floor, his soft, heavily tattooed body shirtless and pale under the industrial lighting. I feel a surge of blood to my face, then my feet, as if my body can't tell if I'm blushing or about to pass out. A man begins to hoist him into the air, the piercings pulling at his skin. I've been suspended by ropes—never just by my own skin—but even I know the vertiginous moment when all that's left of you on the floor are the tips of your toes, and it just seems so unlikely that not having contact on the ground would ever be possible. And then suddenly, he's up.

Twirling above me in khaki cargo pants and brown leather boots, the thin catheter of an insulin pump fluttering from his pocket, the man above the crowd is giddy as hell. His face! He is smiling! Kicking his feet like a little boy and swinging in wide circles over a hushed crowd, he dangles to the sound of nondescript metal music. (Though now that I'm here in the present and thinking back on it, I can't tell you if the crowd was truly quiet in that room, or if my ears were so filled with blood that I was no longer aware of other people.) I thrill at the sight of his taut skin, the impossibility of this. How placid he seems. Keeping him safe and airborne is the rigger, who jumps up into the air to get more purchase on the rope, to pull our dangling man up higher with each bounce. There are four openings on his back, two for each bar. One of them, second one from the left, has started to bleed. Nothing about this man's demeanor suggests this is unpleasant, though hanging by pierced skin clearly involves a significant amount of pain. As his weight swings around the room, the skin around his piercings is changing: bright white half-moons where the flesh is being pulled tighter than I would think possible, the color of red grape skin in the expanse of his upper back, ringing the piercings like a frame. What must he be thinking, with that happy face, that purpling back, all those air bubbles getting trapped under his skin?

My friend, who we'll call CJC, gave me a detailed, graphic description of her own four-point suicide hang. Apparently getting the hooks in her back was the least painful part. After they were in her skin, she went through five minutes of adjustments with the rigger for even weight distribution, and then it was time to fly. The early moments of weight bearing on the hooks were excruciating, and there were very specific sensations coming from her back. "I could feel the skin adjusting and unfolding, the small paperlike sounds from my skin tearing at the entry points, and the feeling the suction being created under my skin as it was being pulled." But with time, the feeling changed.

"Once enough weight was on it, the blood flow was cut, and the skin started to numb and I was left with the feeling of tension." The hardest part, she tells me, was letting go of the ground. "I never knew how much I used the sensation of gravity as a means to orient myself. Whatever I had done in life up until that point was always informed by the feeling of weight, and now I had to give it up. That part doesn't hurt; it's just psychologically daunting, like I was giving up the only true thing I knew. Once I was up in the air I felt the endorphins kick in, the tension from my back had moved to feeling like my chest was compressing. It was both just incredibly blissful and slightly unnerving."

I ask her what she was thinking while she swung in the air, but the feeling is hard to pin down. "I can't really describe where my mind went as it struggled to collate and process the sensations I was feeling. I felt like I had been thrown, but no gravity was telling me that I was coming back."

CJC flew for around thirty minutes, then came down as the piercings became increasingly uncomfortable. Once returned to earth, the hooks felt like sandpaper under her skin. They burned, and her skin felt tight. She had help getting the hooks out. The woman removing the hooks asked CJC if she wanted her to push the air out, the air that had built up under the skin of her back, and CJC agreed. "As she did, I could feel the air sliding around over my muscle and situating itself. It felt like a series of tendons sliding over one another as it slid from cavity to cavity under my skin." Afterward, she tells me it felt like she'd been hit by a truck, but she felt lighter emotionally. Hanging by her back made her feel "generally happier," a feeling that stayed with her for "over a month."

There are expansive parallels between the transcendence of religious penitential pain and all the ways people use the symphonic pain mechanisms of the body to find new ways to be conscious, new things to feel. Having spoken with so many people who have suspended, I find the threads of sacred experience and self-identification looped through their moments in the air. Pain on purpose can be a test, a proving ground. It can be joyful.

As Lee S. described the feeling: "Once I was in the air, I had an almost out-of-body experience. Euphoria flooded my head and body, and there was no more pain." He said that for him, it released a lot of tension and negativity, and that it required him to surrender completely to his surroundings. He did it to explore what it could feel like to relinquish control. "I didn't expect it to be such an eye-opener. I cried a little once I was in the air."

Of course, hanging by hooks in the skin is not some trendy and transcendent darling of the body modification community. Though popular today in communities of high-sensation-seeking pain lovers, the practice comes from the Mandan people of North Dakota.

Specifically, this kind of flesh hanging comes from the Okipa, a four-day religious ceremony that used to be performed each summer, and it was their most powerful religious ceremony. It told of their creation, their sacred origin unfolding through dancers and drummers. Inside their lodge, men fasted, prayed, and suffered. In the *Encyclopedia of the Great Plains* from the University of Nebraska at Lincoln, Leslie V. Tischauser writes, "The younger men generally underwent torture to demonstrate their bravery. Long wooden skewers were pushed through cuts in the skin on their backs or chests, and they were hung by ropes from beams. Their bodies were weighted down with buffalo skulls hung from other skewers thrust into their thighs and calves. The torment was extreme, but crying out was a sign of cowardice, and those best able to stand the pain became Mandan leaders. Women were not allowed inside the lodge, although some would sit on the roof, where they would fast."

The ceremony existed to "unify the Mandans through a ritual of suffering and bloodshed."

In 1832, the artist George Catlin witnessed and painted the sacred ritual, at which time the Okipa had been performed for hundreds of years. It is estimated that the ceremony was likely held for the last time on Fort Berthold Reservation in 1889 or 1890, after which it was suppressed by the United States in a reign of genocidal terror.

Roughly a century later, Allen Falkner, who is now known as the Father of Modern Suspension, was inspired to start a movement based on the sacred practice and began his work to bring secular suspensions to the mainstream. It's important for outsiders to acknowledge and understand the horrifying and traumatic reality that the practice of human suspension was forcibly stopped by white settler colonists only to be later resurrected by outsiders. There is a deeper conversation that must be had regarding this, and as a white person, it will be my job to listen.

These days, people hang in the air for all kinds of reasons, says body piercer and suspension practitioner Brian Belcher. He says that the reasons people do it are purely individual, be it religious, fetish, curiosity, or otherwise. For him, the reasons are physical. "I do it for the excitement of pushing my body's pain threshold to deeper levels, the trance I go into, and the rush that follows." He likens it to a drug, but one without harsh side effects. "Because of suspensions I actually set aside drugs and got the high I needed from suspensions and pulls alone."

I ask him what it's like to hang people in the air by their skin. "Seeing the expression of joy and happiness is the most rewarding part of what I do." As a piercer with nearly two decades of experience, he gets it when it comes to pain and pleasure, having seen the two intersect quite a lot. "Based upon the individual's tolerance level, one may get more excitement from deeper pain levels," he says. "Hurts a little but also feels good." He tells me about one of the best experiences he's ever had, a memorable painful experience of his own at a BDSM convention in Atlanta. "I was doing a chest pull against one of the other guys in our group, when after thirty to forty minutes of pulling I began to have a complete out-of-body body experience." (A chest pull is when you have hooks in your chest, but instead of hanging, you pull against another person.) "I remember pulling harder and harder while trying to rip the hooks out through the flesh of my chest. During the pull, I got down on my knees and began leaning back. At this point my mind started removing my spirit from my flesh. I was hovering over myself looking at myself as in the third person. It was very surreal, and in the video taken my body began to shake, and I went into shock. Once it was all over, and my flesh and spirit became one again, I was in a high for nearly three days. The feeling of pain and the pleasure I received was like none other I'd ever experienced."

Back at the tattoo convention, I am eager to watch the next person to fly. She is tall, slightly vibrating, hyperalert. I think about nuns filled with harrowing resolve, about ascetics and their deprivations, about feeling pain to feel something else, something bigger, something beyond. The sacred ritual of the flagellant, the way the act must have seemed so profane to the mobs who watched all that soiled skin bloom red. The emcee addresses the crowd as the woman makes her way to the ropes. "Guys, it shouldn't be hard for you to cheer for a lady doing something she doesn't want to do," he says, chuckling into the mic. The crowd responds in kind. It is so easy to hate him. Sacred hooks, profane man.

I want to take his scraggly, Kool Aid–red chin beard and shove it into his chuckle-mouth, make him eat it, tape it in there so it gets all stuck in his teeth, but what I am witnessing today is worth so much more than my petty vengeance fantasies. The woman sits on the floor, knees up, as Truitt attaches her to the ropes. Once secured, he tenderly cradles her body from behind, his arms under her arms, supporting her weight as her pierced knees begin to rise above her head. Her bare feet are slim and graceful, her soles covered with the gray patina of dusty arena flooring. As the ropes hoist her up, her spine unfurls to find her upside down, hands gripping Truitt's biceps. It takes forever and only an instant and without a moment's hesitation—who would want to hesitate in the hardest part of this?—they separate. She's up.

Her long red hair hangs beneath her as she dangles by the skin of her knees, her hands clasped neatly behind her. Truitt places his gloved hands on her back and gives her a push, sending her into an arcing swing. She looks angelic, a dreamy smile; closed eyes. I feel high watching her melt like this.

There is an aspect of human suspension and other extreme physical feats of that nature that is wildly compelling to me. I am endlessly drawn to the idea that if you can just get through to some nebulous *other side*, that pain can open up into wild euphoria. Humans play this game all the time, on many different levels, and once you see it, it's hard to ignore.

I think about the graceful body writhing on the ropes before me, about my own wants. There is revulsion mixed with titillating curiosity that flashes in the eyes of people who would never in a million years hang from hooks but are greedy for insight into the *whos* and *whys* of the practice. I do not think there is any one reason a person finds masochism, nor do I think the reasoning is static inside even one individual. Suspended in precarity by the skin above her kneecaps, she twists her body around like a worm on a hook. I watch her and I wonder what joy she must be feeling. What reward could possibly be worth so much pain?

Chapter Four

LAVA MOUTH

Parked in the dusty lot of the county fairgrounds in Auburn, California, I'm in the driver's seat of a rental car, hands imperceptibly trembling. There's a cold bottle of water sweating in the cup holder, a one-use relic of end-stage capitalism that will outlast us all. I'm recording the preamble to my pepper-eating video over and over again, the insides of my lungs thrumming with excitement. (I'm scared.) The seats are deep and dark, and all the lights shining through the icons on the dash are red, making me feel like I am in a spaceship. This feels ominous, appropriate, given the journey I'm about to blast off on. I'm stalling with repeated takes, chipperly talking to my phone over and over, wedging it this way and that in the steering wheel. I'm stalling even now, even writing this, somewhat reticent to fully revisit the sadistic contours of the Carolina Reaper pepper experience I was about to bite into.

It's a wonder, this, the hottest pepper in the world: its green stem between my fingers, its body having all the innocuity of a strawberry, with which it shares both size and hue. Looking at it, I feel the same way as when I hold a tab of acid or a test tube of particularly virulent *Pseudomonas* or a superlatively mean plastic flogger. The potential of a thing, radiating outward from it at a frequency only heard by those with ears tuned to its call. I've taken mushrooms on purpose, and I've taken mushrooms not on purpose, and I must say, I treated the humble object carrying the experience within it with much more reverence when I knew of its Precious Cargo.

I know what this pepper is. I know who grew it: Greg Foster, who holds the world record for most Reaper peppers eaten in one minute (120 grams, which came out to sixteen peppers, if you want to give it a shot). I know that this pepper is hot: I know its Scoville heat unit score (2.2 million, or roughly 600 times hotter than a single jalapeño pepper), and I know what it is going to do to me (very bad things). I've read countless essays about what it's like to eat the hottest pepper in the world, watched a guy on YouTube smoke one in his bong, flipped through episodes of *Hot Ones*, where celebrities like Idris Elba and Paul Rudd choke down molten-hot chicken wings while being interviewed and weeping. I've spoken with world-class hot pepper eaters, read stacks of scientific papers about capsaicin, know that dairy and alcohol can be used to soothe the pain that is coming. I guess I am as ready as I could possibly be.

But I did not bring beer or a milkshake or anything that might uncouple the capsaicin molecules from the receptors in my mouth. Foolish human. I want the full experience.

———

"If you keep them in, no matter what you do, you're going to have, like, the worst night of your life," says the UK Chili Queen, Shahina Waseem. She tells me that out of the seventy-one hot-pepper-eating competitions she's entered, she has failed to throw up afterward eleven times, which meant that she got the dreaded *cap cramps*: gastric distress caused by capsaicin in the GI tract. "You can't get up, you're literally on the floor like palms up, you're begging, and you get cold sweats, hot sweats, everything, and you're literally begging for death." She looks me in the eye, serious as the hell she's describing. "It's that painful. It's like being stabbed multiple times, and it's the worst feeling." After that, she says, she learned her lesson. Whatever happens, she has to throw up. "It burns coming back up as well. But then you think, 'Well, I'd much rather have that than, like, twelve hours of being on the floor, hunkered up, crying and screaming in agony.'"

Welcome to the 2019 Pepper Festival.

Shahina and I are chatting in a small, stand-alone structure on the California state fairgrounds surrounded by volunteers looking for their orders and musicians hiding from the sun, an array of single-serving chip bags, wet beers from the ice tub, and pepper eaters. Competitive pepper eaters. In fact, some of the most highly ranked pepper eaters on the planet. Shahina, one of the headliners for today's main event, sits across from me. She is animated, sharp, striking. All the announcers today will comment on her small stature, as if her lack of bearishness would put her at a disadvantage in this battle of wills. It's something American cis men love to do, act like smallness inherently brings with it a competitive disadvantage, even if that competition has nothing to do with stature. *Sir, this is a chili-eating competition, not a football game.* The Atomik Menace didn't get his name because he's a big dude; he got it downing superhots.

Shahina's nervous. She tells me that rumor has it that her opponent today, Dustin "the Atomik Menace" Johnson, doesn't feel pain in his mouth from chili peppers. A relative newcomer to the scene, he burst onto the international pepper-eating circuit with his win at Ed Currie's Inaugural International Pepper-Eating Contest held a month ago. Ed Currie, for those outside the chili pepper community, is the creator of the Carolina Reaper pepper, which, as I will experience firsthand soon enough, is the current record holder for the hottest pepper in the world.

Dustin, however, refutes the rumor. "I do feel the burn. I honestly couldn't imagine caring at all about spicy food if I didn't," he wrote to me when I asked. "I almost think I'd view it as narcissistic to enter contests without any feeling whatsoever," he says, noting that he is just "relatively gifted that I don't have major reactions." He doesn't, however, usually purge after competitions. "I don't get cramps too much until they're about fully digested," he says, though I cannot tell how much of his stoicism is a well-tuned performance to psych out future competitors. He is quiet in person, well-spoken in writing, and terrifies people in the competitive pepper-eating circuit. Later in the day, Dustin the Stoic and Shahina the Emotive will battle it out at a plastic picnic table, stuffing themselves with superhot peppers, without water or any relief, in front of a cheering, rapt audience who is there to drink cold beer and watch them suffer.

Outside, a small crowd of people mills about the festival. There are food trucks and a stage for the bands; people selling vacation packages, snow cones, and T-shirts; a weed delivery service that can't, technically, deliver in Auburn, which is heartbreaking. Rick Tracewell, the organizer of this event, stands in a white tent before a host of cardboard boxes lined in plastic and bursting with brightly colored peppers. But the main event, for now anyway, is inside the hollow, one-story building that stands low and long by the bounce house: the hot sauce expo.

It's a blessing, this indoor activity, though somewhat ironic that the hottest happenings are coolly shielded from the burning September midday sun. Inside, the walls are lined with the tables of hot sauce vendors giving samples of their wares and chatting with pepperheads. There's everything from honey-sweet pepper jams served on buttery crackers with gobs of cream cheese to concentrated capsaicin tincture hot enough to really make you wail. At one table, as I'm watering from a particularly mean habanero sauce, I ask the man behind the table why he has a sword. He offers to swallow it and let me pull it out of his throat, and when I do it, I can feel his throat muscles twitch around the blade. The person next to him then whips out a full bed of nails and asks me to step on their chest while they lie on it. I decide that I love the pepper festival.

Everyone is here today for one obvious reason: pain—theirs, or someone else's. For now, it's here in this room: couples with unequal capsaicin tolerances sampling salsa and falling into slapstick; people turning crimson and acting nonchalant; people turning crimson and not acting nonchalant. There are few true bystanders, abstaining completely from the peppers and their sauces, but they are still here, witness to the pain contorting so many faces in the crowd. Later, though, the pain will be spectacle at the pepper-eating contest. But regardless of whether the pain is on display or private, here today it's coming from one little molecule: capsaicin.

There are many misconceptions about capsaicin, and for that matter capsaicinoids, a family of molecules of which it is a prominent member. I've heard that it can burn a hole in your throat or your stomach, but that's not true. It also isn't involved in gastric reflux or heartburn. I've heard people say that it can make you throw up blood, or shit blood, or cause otherwise graphic damage, but unless you are downing pure, crystallized capsaicin—which is, in that form, very dangerous—it's just not the case. (I will say that if you have a bleeding gastric ulcer and throw up after eating a pepper, there will be blood in your puke, but the blood won't be the pepper's fault.) Capsaicin allergies are rare and affect fewer than 1 percent of the population, but, like laundry detergent and poison oak and nickel jewelry, capsaicin can cause contact dermatitis in some of those people. For most people, though, eating hot peppers isn't harmful—it just hurts. And amazingly, the burn you feel after a steaming plate of *buldak* or Nashville-style hot chicken has nothing to do with your taste buds.

If we were to anthropomorphize molecules, I would say that capsaicin is a clever little shit. Structurally related to vanillin—the molecule in special orchid pods responsible for vanilla's creamy, iconic flavor—capsaicin is a trickster. It doesn't fuck with the taste receptors in your mouth that alert to things like *sweet*, *sour*, *bitter*, *salt*, and *umami*. No, capsaicin fucks with temperature receptors called TRPV1s.

Specifically, capsaicin is a heat mimic. It activates sensory neurons that alert the brain to the presence of actual heat, not the *flavor* of heat but real, kinetic energy heat. When you eat a very hot pepper, capsaicin binds a specific receptor, the kind that warns your brain when your coffee is too hot. When the competitors today start horking down handfuls of the hottest peppers in the world, their brains will alert to the presence of a molten, dangerous substance—lava, hot coals, actual fire, whatever. But there is no real threat. The game, then, is to sit there and withstand the very real pain without running screaming from the not-so-real danger.

(Mint works similarly, faking the sensation of cold. However, mint and hot peppers act on different receptors. An Altoid and habanero will not cancel each other out; if you eat them at the same time, you will have created inside your mouth an edible Icy Hot arthritis cream.)

How exactly do people determine just how spicy a pepper is? After all, the "world's spiciest" designation is a coveted accolade, and a moving target. As such, peppers are ranked according to the Scoville scale. This scale is used to determine the spiciness of a pepper, which correlates with a number reported in Scoville heat units, or SHU. There are a couple of ways to do this test. The first involves drying the pepper in question, then mixing smaller and smaller quantities of the dried pepper powder in a mixture of water and sugar, until taste testers can no longer detect any spiciness. The degree of dilution required to squelch the heat corresponds with the Scoville score of the pepper.

That was the original and notoriously subjective version of the test, developed by American pharmacist Wilbur Scoville in 1921. Today, we have high-performance liquid chromatography (HPLC), which is a cherry method for separating, identifying, and quantifying the *stuff* of a liquid sample. Using HPLC, researchers can assess capsaicin concentrations without having to rely on the nonstandard and relative sensitivities of the human mouth.

So, what does this scale look like? At the bottom, with a score of 0 SHU, are bell peppers. They have the sweet and verdant flavors of peppers, with none of the heat whatsoever. Poblano peppers, with their mild, emerald flesh come in at around 1,250 SHU. Original Tabasco sauce sits at 3,750 SHU, and jalapeños, as spicy as many people can enjoy, have a rating of around 5,000 to 8,000 SHU.

Climbing up the scale, we hit cayenne peppers at 50,000 SHU. Triple that are habaneros (150,000 SHU). The peppers that the Indian army uses as the base for chili grenades (hand grenades that incapacitate with capsaicin), the bhut jolokia, rank at an astonishing 1,000,000 SHU.

But the hottest pepper in the world, the Carolina Reaper, comes in at a whopping 2.2 million SHU. That's 44,000 percent hotter than a jalapeño pepper, for those who like a little back-of-the-napkin math. Beyond that, pepper spray can get up to around 5.3 million SHU, and pure, unadulterated crystalline capsaicin clocks in at 16 million SHU.

At the pepper festival, I am surrounded by amateur and professional pepper sufferers alike sucking down taster spoons of condiments designed for pain. Every day, around the entire globe, about one-quarter of the entire world's population tucks into meals seasoned to make their mouths hurt at least a little bit, their eyes tearing over steaming dinner plates of phaal curry, buffalo wings, or Som Tam served Thai spicy with vibrant red curls of peppers studded throughout. Hot sauce is booming; by 2021, sales of the spicy slurries in the United States alone are expected to be a $1.65 billion industry. So why on earth do people eat deliberately painful foods?

"Pain can be fun!" Dr. Paul Rozin's raspy voice shimmers with glee. "I think a good case could be made that it's extraordinarily common. You've got a couple of million people doing it a bunch of times a day." Now, unlike Rozin, who says that he is not inclined toward painful stimuli, I'm an avowed masochist; I see masochism everywhere. I am looking for it, I am finding it, I am obsessed with it. As such, one of the foundational studies that has helped me make sense of it comes from Rozin. He is the academic father of the theory of *benign masochism*. When I say "many, many people get pleasure from pain," it's Rozin standing in the wings, nodding sagely. Because it's not just pleasure from the pain itself; *pleasure from pain* can be a pleasure induced by the relief of the pain ceasing. "Some people of course don't enjoy those things, but they enjoy the relief from them, so they'll experience something unpleasant intentionally because the ending of that is very pleasant," he tells me. "That's not the same as enjoying pain. Right?" Or is it?

The question hangs in the air. It's a big one, and it resonates through all of the pages of this book. Upon reflection, I must say, I don't think that the idea of enjoying pain is so black-and-white. If a person deliberately engages with pain, not because they like the sensation of pain but because they like the sensations that come when the pain ends, I think that qualifies as a flavor of masochism. I think this because, anecdotally, and through the years of experiences that I've had relayed to me by hundreds of other people who dabble in deliberate suffering, when people talk about pain on purpose, they almost always talk about what comes next, how they feel after the pain. The dominion over self. The endorphin rush, that hit of homebrew morphine, the lactic acid that makes the muscles tense with a pleasing burn long after the workout has ended. High-sensation-seeking people out there using their bodies to test limits, to feel something wild, to push themselves. There are masochists who are strictly pain-seeking for the sensation of it, but, in my experience, there are so, so many more who use pain as a tool to feel something else. To feel bad to feel better.

In the 2012 paper titled "Glad to Be Sad, and Other Examples of Benign Masochism," Rozin looks at the intersection of pain and pleasure. Or rather, the very broad overlap. In the study, Rozin and his colleagues asked participants to rate twenty-nine innately negative experiences (like sadness, mouth burn, fear, and exhaustion) on a scale of 1 to 100 based on how pleasurable they found them. You read that right: Rozin and his team were asking people how much they enjoyed negative experiences. And they found that about half of the participants *enjoyed* these experiences, rating them at around the midpoint of the enjoyment scale. That is, it is normal and common to enjoy things that feel negative. "Masochists enjoy suffering," he says, "but the kind of suffering they enjoy depends on the kind of masochism they have."

But there's more! Rozin and his colleagues asked participants *when* they got the most enjoyment from negative experiences. And wouldn't you know it, for anywhere from one-quarter to two-thirds of participants, the best part of a negative experience was *the most extreme point they could stand.* That is, for lots of folks (myself included!) the fun in feelin' bad peaks right at the limit of what we can stand. It's fun until it definitely is NOT anymore.

As social psychologist Brock Bastian writes in his book about the role of suffering in happiness, "Benign masochism characterizes the enjoyment of the conflict that arises when these simultaneous positive and negative emotions are activated." Whether you're whipping yourself for Jesus or sex, running marathons for self-esteem or penance, eating spicy food for the taste or the burn, it's all happening on a kind of sliding scale of masochistic engagement. (Careful readers will also note the fallibility of these "or" statements, as if such practices aren't built on a combination of many types of motivations and rewards!)

"A real masochist might actually enjoy pain, which is not threatening to his or her body," Rozin tells me. But who, then, is a *real* masochist? How do you delineate between sexual masochism, benign masochism, and other kinds of pain on purpose?

"I'm not trying to make any huge, overarching declarative statement about the nature of things," I say to Rozin. "I'm just trying to look at this interesting thing from a lot of different angles and get people thinking about the role of pain in their own lives. But I'm not trying to write a self-help book or make a grand theory of masochism." I'm just trying to look at some of the reasons why.

So why, then, do people like spicy food? "I don't think there's one answer to it," says Rozin. "I think benign masochism is a piece of it, but it's also the fact that this experience has been associated with the positive, the people they care about in their life, their parents, their siblings. You know, it's got a lot of positive stuff associated with it. And I don't think there's one account."

I ask Rozin about why so many people report feeling so good after eating peppers, and his response mirrors the dearth of research I have found on the topic. "I don't believe anyone has measured," he says. "You can find endorphin rushes in humans. You can look that up. So it's possible that's what's happening . . ." His voice trails off, then he briefly mentions attempts to study the endorphin rush of a runner's high before coming back to peppers. So, what do people get from eating hot peppers? "You know, we don't know. It's just an area of amazing ignorance."

———

"I tasted a pepper at approximately three thirty this morning." Ed Currie stands behind the counter at his retail store for his pepper empire, Puckerbutt Farms, in Fort Mill, South Carolina. Yes, I said Puckerbutt. Currie eats ultra-hot peppers every day, he tells me. All day. He even puts pepper tincture in his coffee. It's peak season for the farm, and it's a race to harvest all the peppers and fill the orders as they come pouring in, which is why he's been awake since bartenders have been mopping the floors with all the lights turned up. I can't imagine what it would be like to bathe my mouth in capsaicin first thing in the morning. "It's exhilarating," Currie tells me. "The very first one I see that looks like it's going to hurt, and I eat just a bite. But that's enough to knock you out." The mean ones, they are extra bumpy and crenulated, and Currie, a world-class pepper breeder, knows them all too well. And how does he feel afterward? "Oh, I'm an addict in recovery, so I feel pretty good."

It's no coincidence that Currie, the inventor of the world's hottest pepper, is an addict.

"I get a rush. It makes me feel good. It gives me energy." Currie is charming and easy to talk to, brimming with the sly, fidgety intellect of a very clever rabbit, and his eyes twinkle when he talks about his peppers. "So yeah, I'm looking for all those different things."

We move to the back office of the store as his staff bustle about. I am flushed because, prior to the interview, I sampled about six of the hottest hot sauces on offer, including one that was 94 percent Carolina Reaper pepper. It was so thick that you had to dig it out of the bottle with a little plastic spoon. The pungent flames of a thimble full of Reaper paste take me back to the parking lot of the pepper festival, and the heat here, though formidable and making me cry, is nothing like the heat was there. I have a new baseline threshold for pepper pain, forever skewed by my dalliance with the Reaper.

Currie tells me that he was an addict by the time he was a teenager. When he got to college, he discovered hot peppers. "I was looking for a way not to die," he says, also mentioning the antioxidant properties of peppers and their potential ability to help stave off heart disease and cancer. It was only later, after reaching the depths of addiction and clawing his way back out, that Currie began making hot sauce. In fact, he wooed his would-be wife with a jar of peach salsa. Later, it was she who encouraged him to begin selling his sauces, and it was the ladies at church who came up with the description of his wares that would become his company name: Puckerbutt.

Currie says what I've heard many people say: capsaicin "releases a huge amount of endorphins and dopamine into your system. It gives people essentially a runner's high." Though science is regrettably behind in confirming this, I think it's fair to say that, because pain causes an endorphin response, and hot peppers cause pain, by the transitive principle, eating hot peppers probably causes an endorphin rush. It was certainly my experience of them. But as of the writing of this book, we're still not entirely sure. Scientists, I eagerly await your research!

"When you eat superhot peppers, it actually makes you feel good," he tells me. "It gets you high. Like me, I'm a recovering addict. I get high off of peppers, I'll admit it. Sometimes I take too many peppers," he laughs. "You know, I've had to talk to my sponsor about it. But it's not something that's gonna harm me. There's nothing in a pepper that can harm you. You can't burn a hole in anything. You cannot damage, you know, anything beyond repair. It's a sensation."

It's not like Currie is immune to the painful effects of his beloved peppers. He definitely still gets the capsaicin cramps. "There's no way around that," he says. "That's one of the physiological reactions that happens to everyone, no matter what your tolerance is. Cramps, sweating, runny nose, crying, out of breath, spit." He tells me his face is on fire after handling superhots all morning, that he's just sitting there burning. Then he tells me that for him, the peppers are an extension of his faith.

Currie is a devout Christian and shows a deep and earnest gratitude when talking about his faith, in a way that feels welcoming and whole, unlike so many of my personal experiences with religion in the American South. He seems full of reverence for his life, his peppers, his faith, his company, his family. He proudly brags on his staff, many of whom have keys to his house. It's a family affair here, he tells me, blood-related or otherwise. He also giggles like a little boy when talking about the emails people send him describing the pain he's caused them.

I ask him if he likes knowing that people are out there hurting because of his peppers, and he smiles big. "Oh yeah, it makes me laugh all the time." Currie grins at me, delighted. I wonder quietly to myself if this makes Currie a benign sadist.

Later, after a winding conversation about hot pepper breeding and thrill seekers and physiological responses to the fruits of his labor, I ask Currie what he thinks about people using the body's pain response and endorphin system to feel good. Thinking for a moment, he tells me, "I don't understand the people who pierce their bodies and hang, okay? Just don't understand it. Not saying it's wrong, not saying it's bad. I don't understand it. I have no want to do it. But I'm quite sure they're getting the same high I am."

⸺

Back at the chili festival, the air shimmers with heat and anticipation and the inviting steam from the food trucks, tinged with the earthy aroma of gastric distress. A crowd of primarily metalheads, elderly couples, a genre I can only describe as West Coast Country, and sturdy, tan women in their sixties begins to fill the amphitheater and mill about while putting tincture in their beers. Nearby, a man loudly exclaims, "Man, that hurts so good!" There's a Tom Petty cover band wrapping up their set; prior to that, a blues band; prior to that, a Cars cover band.

It's time for the pepper-eating competitions: first, a twelve-person slog through escalating hot peppers, with the winner the last person standing; second, the international battle between Dustin "the Atomik Menace" Johnson and Shahina "the UK Chili Queen" Waseem. Shahina is nervous, jittery, twitchy; Dustin is cool, imposing, stoic. There have been hushed and not-so-hushed whispers all day that there is something *different* about Dustin, not just from her but from others too: maybe some kind of congenital insensitivity to chilis, maybe some kind of bleeding-edge doping (though I can't imagine what). Dustin denies that he's different. Shahina's manager flutters around her, boosting her up; they are conversing quickly and quietly. During the competition, he'll kneel before the table and speak to her from his low vantage on the ground, pushing her and cheering her on. I find out later that he is her husband. She's so nervous that it makes me nervous too. She's never had American chilis, she told me. Meanwhile, Dustin just won a pepper-eating competition at Currie's farm, the very source of the hottest pepper.

We begin with the spectacle of amateurs. There's been a late entry to the competition, Greg Foster, current holder of the world record for most Carolina Reaper peppers eaten in one minute. Earlier, when he gave me a Reaper pepper that he grew himself, I asked him how he got into this. He chuckled like a rumble—Greg is a large, formidable man—and told me that he is an addict in recovery, just like Currie. The first ghost pepper he ever tried caused what he described as a paradigm shift. So, why peppers? "'Cause it's fun, I get high as a kite, and it's a great show."

He said earlier that he probably wouldn't compete today, so it's a pleasure to see that he's changed his mind. In a bit of psychological warfare, Greg sets a quart of milk on the table directly in front of his place setting in view of the other competitors. Competitors are not allowed any fluids during the competition, not even water. The relief offered by fatty dairy products is only allowed after the ordeal is completed, and Greg's milk jug exists to taunt the sufferers.

The first competition consists of twelve competitors, twelve rounds. They are seated Last Supper–style at a long banquet of folding tables on the ground below the stage. They will eat peppers in this order:

Round 1: Jalapeño (8K SHU)
Round 2: Fresno (10K SHU)
Round 3: Hatch (30K SHU)
Round 4: Cayenne (50K SHU)
Round 5: Fatalii (200K SHU)
Round 6: Habanero (200K SHU)
Round 7: Scotch Bonnet (250K SHU)

Round 8: Chocolate Habanero (800K SHU)
Round 9: Ghost (1 million SHU)
Round 10: Trinidad Moruga Scorpion (1.4 million SHU)
Round 11: Carolina Reaper (2.2 million SHU)
Round 12: 7 Pot Primo (1.5 million SHU)

These peppers represent the hottest varieties in the world, from all around the world. But as ubiquitous as chilis are today, it hasn't always been this way. It is hard to imagine savory kozhi curries without Byadagi peppers, or Hunan dishes without Tien Tsin chilis, but in the grand scheme of things chili peppers are a relatively new addition to global cuisine, though they had been domesticated and enjoyed for thousands of years in Mesoamerica. Prior to Christopher Columbus sowing the seeds of Indigenous American genocide in 1492, no one outside of the New World knew of chilis. Among the treasures Columbus was searching for in India, his intended destination, was *black gold*: peppercorns. Instead, he found the Caribbean islands, and, as Jodi Ettenberg reports in her Legal Nomads essay "A Brief History of Chili Peppers," he confused the spicy fruit he found there (*pimiento*) with black pepper (*Pimenta*) and brought *Aji* chili peppers back to Spain with little fanfare. But people quickly recognized the value of the small red fruits, which were a big hit in India and China. And that was enough. The Portuguese began trading the peppers in earnest in the late 1500s, and peppers made their way around the world because of the Portuguese empire.*

Since peppers debuted on the global culinary stage in the 1500s and 1600s, humans have been figuring out how to make them hotter. For example, the star of today's show, Currie's world-record-holding Carolina Reaper pepper, is a breed derived from pepper seeds he was gifted from Pakistan and from the Island of Saint Vincent, in the Bahamas. (The breeding involved a few other peppers as well, which he slyly didn't name.) From their roots in Central America and Mexico, peppers conquered the world and have now returned home as hotter and spicier versions, thanks to the way

---

* Notably, peppers made it to the East Coast of the now United States of America not by coming up the short and direct way from Mexico but through the slave trade. Chili peppers had been such a big hit in parts of Africa that Portuguese slavers brought them along to feed their stolen hostages.

these peppers have been cultivated in the communities that love them, who push the limits of naturally occurring capsaicin. Currie currently has a pepper in development, Pepper X, that's hitting three million SHU, but it hasn't been certified yet. By the time this book comes out, we might have a new world record holder as the hottest pepper on earth. And the competitors before me will be sure to try it.

Right now, the peppers here in Auburn, California, spill out of sheer plastic bags and cardboard boxes, ready to hurt people. By this point in the day, I know what the competitors are in for because I have already tasted the hottest pepper in the world. I watch with lurid fascination.

The first few peppers go down relatively easy. The pros Shahina and Dustin hand out the little hate gems to the willing twelve; no one is wearing gloves. Pale, flip-flopped feet jiggle under the table; a man adjusts his sweat bandana. As the peppers get hotter, people start to rock back and forth. Some of the twelve look concerned, others angry, others merely annoyed. The crowd stares with rapt attention, the circus of suffering unfolding just as anticipated. The tension! The willpower! This contest has all the drama of combat sports, with less brain damage. Humans love to watch other humans suffer—football, boxing, MMA, rodeo, ballet, *America's Funniest Home Videos*, *Jackass*, horror movies, even just the potential for a car crash at a NASCAR race is enough to draw viewership—and boy howdy does a pepper-eating competition offer a gentle, safe way to indulge in such voyeurism.

It's also fucking impressive. It's not a passive pain we're watching; that is unappealing to me. We are watching active, chosen pain. These people can stop at any time. Their suffering is not outside of their control, ergo I can watch it with moral impunity and enjoy the dramatic narrative of adversity and triumph. Or failure, as it happens. People start dropping out. There's no milk or water or anything at the ready for them; a sadistic lack of planning and aftercare greets them in their defeat.

Others are quitting. Greg Foster takes his lone jug of milk and storms off in a huff. It gets down to five, to four, to three. A man rushes by me and vomits in the grass near my feet, his pants baggy around his plumber's coin slot that's now in full view of the audience. He's hurting, bad. Eyes red, chest heaving, he shakes his head back and forth with great sadness, looking seasick on his feet. The second-place finisher, the penultimate quitter before the victor, also flies by me to barf in the grass. He is sweaty, face wet, grimacing in the California sun. It smells like stomach acid and copper, sharp and pointy in my nostrils. There's no water around, no milk, no ice cream. A small, middle-aged military veteran sees this and runs off in an attempt to find remedy. I give the men my beer, the only liquid in the immediate vicinity, and then there's more loud retching.

The hero of the hour returns with ice cream, and as the worst of it passes, the laughter starts. Easy conversation flows by the puke piles, as Shahina and Dustin begin their head-to-head contest. The men around me talk with the rapid flow that comes with one or two lines of cocaine. The *whooshing* feeling that starts to build when you start to come up on some speedy drugs—that's what it feels like I'm watching. Earlier they were quiet and, as many mentioned to me, nervous. But now the words come fast and easy, flecks of partially digested pepper clinging to chins and T-shirts alike.

As they eat in silence, Dustin sits next to Shahina, twiddling a Rubik's cube. Shahina is already swaying a little bit, her slender shoulders broadcasting the nightmare unfolding inside the seams of her digestive system. She told me she had no doubt that she could make it to the tie-breaker round with Dustin, confident in her ability to make her body choke down any pepper. She opens and closes her mouth like a goldfish, or someone on an uncomfortable amount of meth. She says that when she eats these peppers, "everything literally is on fire in my mouth. You know when you burn a bit of skin on the oven or something and then you're like, 'I've got to air it! I've got to air it!'? That's what happens to my mouth. I try to get air in to cool it down." I like how unrepentant she is about the spectacle. "People are like, 'Look at her, she has no etiquette. She's eating with her mouth open.' I'm like, no, I literally can't close my mouth, it's so painful." She says her worst fear is not being able to get the peppers back up again.

Shahina is relentless in her pursuit of mastery, and prior to the competition we talked at length about her almost-unblemished record of 71–1. The one loss, she told me, was in a speed-eating contest, and the peppers weren't even very spicy. She recounted the scenario in grueling, animated detail. She is *bothered* by this loss, and her impassioned description of the event grants her the title of one of the most competitive people I have ever met in my life.

Shahina and Dustin eat all the peppers given to them, stacked in little plastic Tupperware containers, the kind relatives send you home with leftovers in. They eat all the peppers without stopping, without water, without dropping out. At some point, they drink hot sauce. A drummer has appeared onstage, giving ritual drumrolls to underscore the hellscape unfolding before us at a lone folding table. Cameras hover about, people scream for their chosen contender, and my god, does Shahina look rough as hell.

It's time for the tie-breaker. If no one taps out due to heat, which they haven't, then the two have to enter a superhot speed-eating competition. It is this part that Shahina was most nervous about. In the past, she has had trouble willing her hand to her mouth. She gets to a point where it's too hot, and her body starts to literally and physically reject her efforts. "Your body just refuses," she said. "It's like opposite ends of a magnet."

Panting and shuddering, they start the speed round. Shahina mows through peppers at alarming speed. She is not having a good time, and it shows, but the crowd is screaming, and she just keeps shoving peppers into her mouth with both hands, monomaniacal.

But the Atomik Menace is finally cracking. His solid, fleshy face has become pink with effort, eyes closed, stoicism slipping away like it's melting in the heat. He's still fast, chewing hard despite what looks to be vastly unpleasant distress. The drums are beating, the sun is going down, and the crowd gets louder and louder, screaming us into the end of this wild, pepper-fueled inferno. Shahina's husband is on his knees in front of her, hands pressed onto his nervous thighs, staring at her intently, shoulders clenched. It's so close, and the peppers run out at the same time, both of them with a mouth full of raw vegetable matter and molten lava, a brain screaming RUN and FIRE and NO. Everyone is sweaty and loud, people on their feet, the spectacle reaching its apex.

Shahina swallows hard, opens her mouth wide, and it's over. She's won! The crowd goes wild; people are screaming hard. She and Dustin both look wrecked, but it's over, and there's milk to be had and barfing to attend to. The once-riveted crowd disperses quickly, everyone trying to beat the traffic jam that's about to happen in the parking lot. There's puke at my feet, and I watch Shahina tremble like a little mouse, victorious. When I congratulate her, she is dazed, distant, wincing, but nevertheless awash in the sweet dopamine bath of victory. It's hard for me to comprehend what she's feeling, how far away she must be from this table, this festival, California, the planet Earth in general. But I have an inkling.

Let's back up two hours.

In the midafternoon hours before the pepper-eating contest, I have sequestered myself in the solitude of my rental car. For hours, I'd carried around my little wrinkly harbinger of fiery doom, the Carolina Reaper pepper that Greg Foster gave me, in a clear plastic baggie. The plastic bag materialized after festival organizer Rick saw me carrying My Precious barehanded and cautioned me—no, admonished me—to be careful touching it, lest the oils wind up in unfortunate places. The person next to him, who was selling handfuls of peppers to ambling festival-goers, vocally agreed with him before announcing that they did not want to work without gloves.

In the car, I take a posed shot of the Doom Pepper in its little makeshift hazmat suit leaning up against my kid's Snorlax plushie and shudder violently. It is somewhat horrifying to place such a sadistic object near where my sweet baby rests her sweet face at night. But she sends Snorlax with me on reporting trips to keep me company, and this was the kind of scandalous photo op that I knew she'd love.

After recording the experience on my phone, I realized I was going to have to come up with a more reader-friendly way to describe the sensation than twelve straight pages of the word *NO*, repeated in different fonts and sizes, over and over and over again for as many pages as it takes to read only the word *NO* for about forty minutes, the approximate duration of my pepper journey. This is my attempt to use language to impart to you the sensation of eating the world's hottest pepper, but please know that, in my heart, I am screaming *no no no no no* the entire time you are reading this. Until the part where my brain starts screaming *yes*.

Like a dog unsure whether they are really allowed the treat on offer, I open my mouth, bring the pepper to my teeth, balk, pull back, then go in for it. I bite almost the entirety of the pepper off and begin to chew, returning to nibble the rest of its red flesh from its rough green calyx. Prior to eating the pepper, I decided that I was going to chew and spit it, lest I endure the capsaicin cramps on my red-eye flight back to the East Coast. Even in my most adventurous state of mind, the idea of enduring six hours of paralytic intestinal cramps while trapped in an airborne shit box was too much for me. The world's most ferocious pepper-eating competitors took steps to avoid getting the "cap cramps." I decided I would too.

But the decision to spit the pepper out meant that I was doomed to overcompensate elsewhere. It's only fair, right? And I, stubborn, overly confident idiot that I was on that ninety-degree day in sunny Auburn, California, decided to atone for my plans to spit by absolutely liquifying the pepper with my big, fat molars. Sitting in the car, stomach pressed urgently up against my pyloric sphincter, lungs protesting the mere vapors of this fresh horror unfolding in my oral cavity, the thought occurs that I might be chewing the pepper too much.

"They taste good!" I chirp into the camera, every ounce the brown-nosing teacher's pet I've always been, red soup sloshing through my mouth like Dante's gazpacho.

The pepper tastes good for about half a second, but then it does NOT taste good. At the beginning, it was sweet and sharp and almost delicate, but then it turned into something else entirely and now I am having very strong feelings about it. Telling the camera that the pepper tastes good is mostly magical thinking, something I'd heard other pepperheads say, as if proving that their mastery over heat allowed them the godlike powers of appreciating the finer tasting notes of fucking lava.

I spit the pepper onto the ground and close the car door with a dramatic exhale, then burst out in riotous laughter, cackling in the face of my obvious mistake. A gasp catches hold of me and hot pepper spit hits the back of my throat. An ominous coughing fit erupts immediately, and my first thought is a very quiet, very stern, very unambiguous *Oh shit*.

There is a siren in my body and it's coming from my mouth and I know, in my heart of hearts, because I did my research, because everyone has told me this, because the sensation is already escalating, fast—I know that the pain is only beginning. Laughter turns to ice-cold concern and I get serious, immediately. I lean out the side of my car and silently retch hard, spraying saliva onto the fist-sized blot of crushed pepper seeds and vermillion ooze that sits atop the baked dust parking lot. I close the door, take my glasses off (which I should not have done), and wipe pepper spit from my face with my bare fingers (foreshadowing!), then clamber again for the door handle.

I lean urgently over the ground, abdominal muscles clenching in revolt as molten-hot spit streams from under my tongue and behind my teeth as if a dam inside me has broken.

I sit back in the car and try to take an accounting of my situation. I have never felt regret so clear, so unquestioned, or so sudden. The camera is still rolling, and I try to remain calm, but I know that any pretense to decorum is over, confident in the knowledge that the things this pepper is about to do to me are going to be unfathomable because, empirically, they already are. And we are just getting started.

I scramble for the cold water I brought along. Fuck the ride, fuck staying on this ride, fuck this feeling in my mouth. I breathe and a dribble of wicked spit hits the back of my throat and my lungs contract in a cough that feels like a threat and a wheeze and a desperate plea for none of this, *Oh god, none of this please, please no, please no.* My intentions were to talk to the camera through the whole ride, but it rapidly becomes patently clear that that ain't fucking happening right now. I start to squat in my seat like some kind of weeping gremlin, drawing my feet up under me in a desperate plea to what? Move away from my mouth? My mouth, which feels like it is actually, literally full of coals. That's not an exaggeration and, yet, it still gets worse!

I perch, a crying gargoyle, snot everywhere. Water isn't helping, so I take a sip of a hot, flat kombucha that has been baking in the car all day. This somehow makes the pain WORSE and FULL OF NEEDLES and I shake my head immediately, as if I could dislodge the experience before it embeds as a memory.

Despite having ardently sworn to avoid this very act, I take my fingers and put them directly, fully, and wholly into my eyes. I do not realize what I have done. It is true slapstick, the earnest performance of a previously considered bad decision. I told myself not to do that, and the second I wanted to, I did it anyway.

Because of this, my left eye and its surrounding area learns to speak a whole new language composed solely of ways to say FUCK YOU as loudly as possible, as quickly as possible, without ceasing.

The heat in my mouth is climbing and everything in my face is screaming. I start doing Lamaze breathing; I am coming unraveled. I take a sip of cold water and it brings the horrible heat into my throat but does nothing,

NOTHING. NOTHING. I take another sip. A different bad sensation is better than the same bad sensation, I reason, because I would do anything to get away from the literal portal to hell unfolding in my mouth. Anything. Come here and punch me in the face so hard that when I wake up, there is peace.

How will I describe this?????? my idiot brain wails, screaming like an infant. I begin to rock my body back and forth. I am lost in this, the heat pouring through my mouth and down my throat. I try to breathe, try to think, but I can't. I am losing to this. I am losing myself here. I am slipping. I try to talk to the camera, but I can't, try to make sure my face is in the frame, but why? I am sliding out of the reality that contains this camera, the parking lot attendant, reasonable social decorum.

My eyes glaze into the distance. After retreating into inner space with eyes closed, fingers pressed, I feel it coming. It's getting worse, and I start quiet, deliberate bargaining with myself, as if I could think my way out of this.

This will end, just breathe through it.

You've given birth before, which took a lot longer and was worse.

This will end.

The sensation feels unsurvivable; it feels scary and dangerous and infinite. I know that this whole ordeal will take about forty minutes, and then it will be over, and I hold that truth in my brain like a tether to reality. Just forty minutes. But I am only five minutes in and riding out the rest of this seems unimaginable.

The pain is getting bad. I lie down. I think about it, think about myself. I think about the pain, and I think about fire and I think about my options. And so, at some point, I decide to do what I came here to do. You can see this moment in the video, the eerie transition: I cough, settle. My mouth curls slightly into a deranged smile and I say it out loud.

"All right."

And so it begins.

It is my humble opinion that, in a case like this, at some point, you have to decide to really fucking just take the ride. On the video, I am leaving for somewhere else. Panting, eyes cloudy. I have all the equipment I need to get through this. I know it in my bones.

There is an interiority that comes during pain, a way to burrow into the body, mind, same thing. I breathe deep and hard into myself, this moment. My lizard brain, my supportive body, my sturdy self. Ultrarunner Courtney Dauwalter has a phrase that describes the way she suffers during the hundreds of miles she runs: she calls it the pain cave. I like the image. Pain creates walls, a container of sorts, sealing you off from the world. But there is an opening to that cave, a light in that fucking tunnel, an end to this. And it's coming.

Moving my body feels good, the rocking motion focusing my mind on the task at hand. The task, you see, is simply to wait it out. I am here to ride this fucking idiot ride I got on and see it to the end. Bold choice, as it's not getting better yet! In fact, the pain has been getting steadily worse, but you know what? This is where I am, and I am going to fucking reckon with it. By the end of things, it will have taken about the forty minutes I was promised.

I squat in the car seat again, rising into myself. Spit runs freely from my mouth onto the leather seats, and I am crawling out of this pain cave by making myself at home within it like an uninvited guest. I breathe into the fire in my mouth; I think about it. I focus my brain on how it feels inside my chest, my pelvis, what my shoulders are doing. I think about my mouth and I think about nothing at all and I think about that guy who free-climbed nine-hundred-meter El Capitan without a rope and how I get scared on the kids' climbing wall at the trampoline park. I think about how I wish my partner was here. I think about the word *NO*, but also on fire, and I think about how the world is ending and that I have a plastic bottle in my rental car and then I feel a wave of spit crest my lip and I notice my mouth and I rush into it.

I'm dissolving into myself, this pepper, my face. I want out and I want through. My eye is bright red and shrieking and it's not even the worst thing happening to my body. I know this is fucking fake! I know it, I know it, I know it! I know that I am not in danger and that this pepper will cause me no harm, and none of that matters because I am at the mercy of the electric tofu between my ears.

So, with that, I surrender. I am gone into the fire.

I clench my fists and pant and snarl, and I root down to the base of the screaming in my mouth. I bear down and I take it. I take it and I like it and

I start to fucking smile like a psychopath because I know, in that moment, I know that I'm going to fucking ride this monster all the way to the end. It's my fucking monster and this is my fucking party.

And moments later, when it starts, it starts in the part of my stomach where butterflies are felt, that exact spot. It comes and it surges through my body too fast. It knocks me stupid and I scream and crumple into my seat, sliding onto my back and clutching my chest. It feels like a hot bath, good drugs, the color pink but also the color red, having a nosebleed directly onto a puppy, coming to after fainting, and being electrocuted with a cattle prod but in a good way. It feels like all of these things at the same time, and I rub my throat and ready myself to accept the true gift of the Carolina Reaper pepper.

My body's pain response floods my system and I am on fire and I am fucking incapacitated with joy. I laugh and shriek and shake my head back and forth hard enough to rattle my teeth. I writhe and shiver and thrash about. I soak this up like a sponge and I wring it into my mouth and I gargle it with relish. I am elated, victorious, lost—miles from the cave, alone in the stratosphere. I fully feel like I am rolling and I wiggle in my seat, unable to conceal my joy with stillness. It is rapturous. The pleasure comes in cresting waves, and I stand in the surf and let them crash into my triumphant body.

My breathing slows.

I wallow into myself, lingering in the animal recesses of my body. Skin damp, hair matted, lips swollen. I rest my body in it. And when I am finished, I climb back out of the clouds and into the dusty parking lot. I am happy.

The first thing I see, upon returning to earth and leaving my car, is a remote-control electric car, with nary a controller in sight. Atop it, snuggled next to a frozen bottle of water and perched on a regal pillow, is a small bunny rabbit, wearing a pair of round, purple sunglasses. The toy vanity tag says "NUGGET" and I have to stop and take a picture, just to make sure that any of this is real.

Chapter Five

THE NAME OF THINGS

IN 2009, AN ASPIRING WRITER POSTED A NOVEL'S WORTH OF STEAMY, Twilight-based drama to FanFiction.net, under the nom de plume Snowqueens Icedragon. For those of you not familiar, fan fiction (or fanfic) is a massive field of online literature where fans contribute to their favorite fictional universes with all manner of writing, much of it dripping with sexual content not available in the originals. Snowqueens Icedragon's work was an explicit retelling of Stephanie Meyer's teenage vampire smash. It featured Bella Swan and her ageless paramour, Edward Cullen, in a variety of X-rated scenarios. It was, in no uncertain terms, a wild success, becoming a widely read companion to Meyer's more chaste saga of longing, denial, and romance. But success in such a forum was not to last.

By the time the story was booted by the fan fiction megasite for its sexual content—which is technically verboten, though the rule is widely flouted— the titillating story of BDSM *Twilight* had already racked up over thirty-seven thousand reader reviews. With the winds of a new fan base now filling her sails, the author moved the story over to her personal website. Soon after, she performed a little trick that's known in publishing as filing off the serial numbers, removing from the manuscript all references to her sexy, hematophagous source material. With a few name changes, the work became legally ready to sell on the open market: Bella became Anastasia, Edward became Christian, and Snowqueens Icedragon's "Masters of the Universe" became E. L. James's Fifty Shades of Grey trilogy.

You probably know where I'm going with this.

Erika Leonard, the London television executive behind the now ubiquitous pen name, was about to get a rather large pay raise. From 2010 to 2019, the first book in the series sold an astounding 15.2 million copies in the United States, making it the best-selling book of the decade. And it wasn't just the initial, eponymous book that did well: slots two and three on the decade's top 10 list were filled by the second and third books in the series, *Fifty Shades Darker* and *Fifty Shades Freed*, respectively. The NPD Group, a market research company, reported that the trilogy sold nearly thirty-five million print and e-books in the United States, in that decade alone. On its home turf of the United Kingdom, *Fifty Shades of Grey* became the fastest-selling adult novel of all time. It has been translated into fifty-two languages. There have been three major motion picture adaptations of the books, which were commercially successful, though, like their borrowed source material, near universally panned by movie critics. RottenTomatoes.com, a movie review aggregator that assigns meta scores based on multiple reviews, gave *Fifty Shades of Grey* a dismal "25% Fresh" rating from the critics. The other two movies? The second scored an impossibly low 11 percent, the last sliding in at 12 percent.

It wasn't just the movies that were despised: the critics absolutely loathed the books too, saying that "James writes as though she's late for a meeting with a sex scene" (Zoe Williams, *The Guardian*), and "compared to the other incredibly problematic aspects, the poor writing in *Fifty Shades* is almost welcome" (Marama Whyte, Hypable), and "if Jane Austen . . . came back to life and read this book, she would kill herself" (Dave Barry, *Time* magazine).

But the readers? The readers ate it up. Society was coming into the age of the e-book, which made it easier for people to stealthily read the BDSM fantasy that, suddenly, everyone was talking about. (Some people even credit the sultry series with propelling the sales of e-book readers so people could read their smut anonymously.) By 2020, James had an estimated net worth of around $150 million.

I bring this up to remind you that humans love, love, love reading about sex. *Fifty Shades* hit a sweet spot of titillation, Cinderella story corniness, wealth fetish porn, and predigested, quick-reading prose. It took off like fire through sawdust. In spite of significant criticisms that the book misunderstands BDSM as abuse, a lot of people read those fucking books.

This is of great interest to me personally because it demonstrates broad public interest in a very common practice, one that modern Western culture typically considers to be taboo: pain for pleasure. In the book, Ana and Christian explore common tropes of BDSM literature, like relationship contracts, impact play, and the feeling of being overwhelmed by a controlling paramour. With the success of these novels in mind, I ask you to take these little publishing statistics and glimpses into popular desires and place them carefully on a shelf in your memory, just for a moment. We are going back to the Victorian era to witness the runaway success of some other books about kinky sex. I'm going to show you something.

In 1836, a boy with quite an outré future ahead of him was born in what is now Lviv, Ukraine, but at the time was a territory of the Austrian Empire called the Kingdom of Galicia. The father, Leopold Johann Nepomuk Ritter von Sacher, was an Austrian civil servant who later went on to become a commissioner of the Imperial Police Force in Lemberg, where they lived. The mother, Charlotte von Masoch, was a Ukrainian noblewoman, and, because of an accident that claimed the life of her brother, she was the last of her royal line. A formidable woman, she convinced her husband to join their names in order to preserve her family lineage. Leopold senior acquiesced. Had he not, you would probably be reading a book about sacherism right now.

Leopold the younger was formed under the redoubtable presence of his parents and was greatly influenced by his beloved nursemaid, a Ukrainian peasant woman named Handzya, who largely raised him. Though he was fond of reading explicit biblical stories of suffering saints and mutilated martyrs, Handzya also regaled Leopold with evocative folktales, stories that often featured cruel, powerful women and elaborate tortures. The settings and characters were usually Slavic, like Handzya. Leopold adored his nanny and later wrote in his journal that he "owed her his soul."

It was a period of high tensions across the continent. Ten-year-old Leopold and his family fled a peasant revolt in his home city, an event called the Galician Slaughter. In this uprising against serfism, over a thousand aristocrats, landlords, and persons of nobility were murdered, some of them even losing their heads in the street. The imperial response was equally brutal. During this time, baby tween Leo got the hots for his cousin Miroslawa, who impressed him in a jacket trimmed with white fur, her two guns and a dagger dangling from her waist. "On the barricade she was a beautiful Amazon, rifle in hand," he later recalled. Some people's childhood fantasies evaporate over time, but not Leopold's. His fixation on powerful, violent women would persist throughout his life, crystallizing into a lodestone that would draw him ever onward in his quest for his ideal.

Later, a trip to the Imperial Museum found teenaged Leopold face-to-face with a painting that would change his life. A painting that led to an obsession, that led to a widely celebrated book, that led to a 1967 Velvet Underground hit song (that led to me winning second place in a Hunky Jesus contest with a stigmata nun burlesque act; it's a wild world out there). The painting that made such a profound impact on this eager teen was by Dutch painter Peter Paul Rubens. For those trying to place that name, Paul Reubens plays Pee-wee Herman, but the painter Rubens was (and is) the biggest name in the Flemish Baroque movement, melding the aesthetics of Italian Renaissance and Flemish realism for lush, overflowing paintings and dramatically lit flesh. The painting that young Leopold fixated upon was a portrait of the painter's wife, Helena Fourment. In it, she is luminous and nude but for the furs that drape her. Furs would drape themselves over Leopold's imagination for the rest of his life. The book he would go on to write, you might have guessed, is *Venus in Furs*.

By all accounts an exceptionally bright student, Leopold flourished academically and was very diligent in school, studying law in Prague in 1852 at the age of sixteen. He was a big history buff who reveled in tales of torture, martyrs, pillages, and orgies. As his biographer James Cleugh notes in his investigation, "It was now also Leopold found in the *Confessions* of Rousseau so close a correlative between the fascinations of physical torment and the irritations of concupiscence." Rousseau was a Swiss philosopher and composer from the previous century who commanded great respect in academic circles and who made no secret of his enjoyment of being spanked by a dominant mother figure. Leopold, in the annals of history at least, had found a kindred spirit.

In 1856, Professor Leopold Sacher-Masoch, now twenty, was lecturing at the Karl Franz University of Graz. He showed promise as an academic but was criticized for making overly dramatic, sweeping statements, which absolutely tracks with what a fussy, dramatic character he was. This police commissioner's son was clearly rebelling against his father's crisp mustache generation. Leopold junior always seemed to be in some state of elegant but unkempt dress, with beautiful, too-long hair and a penchant for poetry and sighs full of longing. It was never his lot to adhere to the dictates of the stuffier academics of his day. So he began writing fiction and after three years was making enough money to resign his academic post and go full-time. With recent advances in printing technologies and rapidly expanding literacy on the Continent, his timing couldn't have been better. Like a certain London television executive 165 years later, Sacher-Masoch quit his safe day job to become a saucy writer.

What exactly did he write about? Well, lots of things, really. He wrote colorful, often humorous ethnographic accounts of Slavs and Jews that earned him great respect. His fight against rising anti-Semitism lasted his entire life and made him many friends in Jewish communities. He railed eloquently against wealthy elites (while enjoying all the privileges as the son of wealthy elites) and the destructiveness of war.

However.

What Sacher-Masoch really became known for were his tales of powerful, cruel women. They were almost always of Slavic origin and dominated his work from the very beginning. His biographer writes that in fiction "one could invent one's own ideal beings, hard-headed but passionate noblewomen, used to authority and fond of it to excess, who stopped at nothing to satisfy their lust for dominion and symbolised it by never leaving off their furs and always having a dog-whip, a horse-whip or a knout within easy reach." Judging by the tremendous sales of his books, lots of readers shared Sacher-Masoch's interests. People just really love reading about kinky sex.

Sacher-Masoch's body of work is extensive and varied, but there is no denying that *Venus in Furs* launched his career and raised his personal notoriety to another level entirely. Published in 1870, the novella follows the brief and tumultuous affair of deeply romantic Severin von Kusiemski and free-spirited but playfully cynical Wanda von Dunajew. A lot of literary analysis boils the protagonists down to a mushy pulp of one-dimensional characters, but these two are actually quite complex and conflicted. Severin claims to want an ideal, chaste wife—caring, kind, and matronly, just like his dear old mama. (Young Leopold didn't even try to obscure his mommy issues.) However, Severin seems fated to never find this ideal, so he invents another, the evil mirror image of the original—a brutal "tsarina" who will treat him like an animal, like property, who will destroy him utterly if she wishes. Severin is a man obsessed, burning with ethereal visions and grasping eternally at ideals that the material world never manifests for him. Wanda von Dunajew, on the other hand, is out to have a ball:

"No, I forgo nothing, I love any man who appeals to me and I make any man who loves me happy. . . . I am rich, and beautiful, and just as I am, I live cheerfully for pleasure and enjoyment."

Wanda doesn't live to serve one man nor to destroy many, but Severin is a captivating speaker. He convinces her to try on the ermine mantle of the conqueror for a little while. She lives for experience, and this is certainly a new kind of experience for the lovely libertine. They travel by train to Italy, with Wanda playing her role of cruel, wealthy debutante. Severin poses as her manservant, "Gregor." He rides third class and fetches her bags for her. You can imagine how exciting this kind of fantasy might be for the author, son of a baroness and an imperial police inspector. The lovers sign an explicit slave contract that grants all powers, spiritual and corporeal, over "Gregor," including the right to murder him if she feels so moved. There are explicit beating scenes that would make a Fifty Shades fan blanch. Wanda takes a strapping Greek lover, and together they beat and degrade Severin further. It would be tempting to say that everything's going Severin's way, considering this is exactly what he had been clamoring for, but it's more complicated than that. He is truly tormented by the torments he's brought upon himself. He can barely contain the jealousy and humiliation burning within him: "The sensation of being whipped before the eyes of a woman one adores by a successful rival is quite indescribable; I was dying of shame and despair."

"How creative!" you might say. "What an inventive writer!" The thing is, *Venus in Furs* is almost entirely autobiographical. Sacher-Masoch took that trip to Italy. He posed as "Gregor," the footman to his lover, one Fanny Pistor Bogdanoff. They dabbled in furs and whips and even took on an Italian actor to play the role of strapping interloper. Prior to Fanny, the dashing young writer had taken up with a married woman in Graz and had convinced her to have a tryst with a bogus "Polish count" who ended up actually being a Russian chemist shop clerk on the lam for robbing the till at his work. It turns out "the Count" had syphilis and gave it to Leopold's lover, so he ditched her, pretending to be appalled by her treachery. The guy was a fucking mess.

A little later in life, Sacher-Masoch married Aurora Rümelin and spent a decade coercing her into similar scenarios. They also signed a slave contract. He took out personal ads for her to find lovers and refused to continue writing and supporting their family if she did not become the "Wanda" of his fantasies. A woman of that time did not have many options to support herself, especially once she was married and had children. Aurora gave up her name, her former self, and donned the symbolic furs her husband coveted. According to her *Confessions of Wanda von Sacher-Masoch*, "From then on, not a day passed without my whipping my husband, without proving to him that I was keeping my part of the contract. At the beginning my repugnance was great, but little by little I became used to it, although I never did so other than reluctantly. Seeing that I was doing as he wished, Leopold set out to make the experience as painful as possible. He had whips especially made to order, including one with six lashes studded with sharp nails."

Studded with nails! Leopold was not fucking around. He is what people in the BDSM scene today would call a serious bottom. He wanted pain; he wanted degradation. And neither one in small measures. And yet, like Severin of *Venus in Furs* fame, he was always conflicted, never satisfied more than momentarily. The gulf between his mystical ideals and real, fleshy human beings caused Sacher-Masoch and his loved ones endless distress.

And a hot new source of distress was just around the corner.

The dusty annals of history are overflowing with the names of presidents, emperors, battle sites, and sunken ships, but it's only once in a blue moon that you come across a name like Richard Fridolin Joseph Freiherr Krafft von Festenburg auf Frohnberg gennant von Ebing. For the sake of expediency and saving on printing costs, we'll call him Dr. Richard von Krafft-Ebing. Late nineteenth-century Austria, where Krafft-Ebing began earning his place in said annals of history, was an exciting age of discovery and new technologies. But there was still plenty of archaic pomp left to accommodate a title-laden name like his.

Austria was actually the freshly minted Austro-Hungarian Empire. European kingdoms were shifting around like hyperactive tectonic plates, but the ancient powers at the top, such as the Hapsburgs and the Roman Catholic Church, weren't quite ready to hand over the reins to modern democracies. It was a dazzling age of steam trains and electric streetlights, but generals still donned tall, feathery caps as they rode horses into battle. Revolts sprouted everywhere like discs of mold on a cheese rind. The Spring of Nations in 1848 saw major uprisings in over fifty countries. New maps needed to be printed annually. It was a full-time job just keeping track of all the new names of kingdoms and provinces.

For all their fickleness, names, as Richard von Krafft-Ebing knew very well, are rather useful things. They allow us to distill complex patterns of concrete traits and abstract ideas down to a few syllables (or in the good Herr Doktor's case, a few dozen syllables). Names are then available to us for repeated use and further analysis. And analysis, as it turns out, was Krafft-Ebing's bailiwick. He and a handful of others were pioneering boldly into new territory in the field of modern psychiatry. It's hard to imagine a time before humans asked other humans about their childhoods, fears, ambitions, and shrieking night terrors, for money at least, but in fact you can roughly trace the current modern practice to this era.

To be fair, people have been exploring the wild expanses of the human mind for at least as long as we have records of humanity. These explorations took place in the fields of philosophy, poetry, history, mythology. History is full of fascinating musings, brilliant insights, and staggering inaccuracies regarding the human mind. Egyptian papyrus scrolls dating from the sixteenth century BCE discussed depression and dementia. Ancient Chinese and Indian practitioners described the steps to achieving a balanced, healthy soul in great detail, blending observation, herbology, religion, and magic. The medieval Arab world dedicated very advanced wards to housing and caring for people with mental illnesses and composed impressive documents classifying numerous known conditions. However, at the same time, Europe was plagued with Catholic superstitions and a tragic lack of useful science. Many people suffering from mental illnesses were condemned to imprisonment, torture, and death for being possessed or having satanic affiliations.

But during the Enlightenment, the way Westerners thought of and pursued science underwent a revolution as intense as the ones convulsing the governments that ruled them. (Fun fact: The term *scientist* wasn't even coined until 1833. Before that, you'd have to contact your local "natural philosopher" to get the latest scoop on luminiferous ether or phlegmatic humors.) It wasn't until the eighteenth century that pioneers like Philippe Pinel and William Tuke, in France and England, respectively, dragged European mental healthcare kicking and screaming into a more compassionate era. The fields of modern psychiatry and psychology really began to coalesce around a network of mid-nineteenth-century thinkers, many of whom were concentrated in France, Germany, and Austria.

It was in this rambunctious milieu, with its burgeoning scientific communities, full of thrilling new trajectories grappling with outdated practices, that our generously syllabic doctor found himself, fresh medical degree in hand, wandering through the sea of red-tiled roofs of Graz, Austria. Originally from Germany, he had been selected in 1872 for a professorship at the university in Graz and had been appointed as the medical superintendent of the nearby Feldhof mental asylum. Feldhof resembled an oubliette more than a place of healing, and the doctor would spend years unsuccessfully agitating for reform there.

Never one to accept defeat or even to slow down, our industrious doctor kept his nose to the grindstone, cranking out esteemed psychiatric texts at an alarming rate and earning quite the name for himself. "By 1870, Krafft-Ebing was the best trained neuropsychiatrist on the continent," according to doctor and translator Franklin S. Klaf. Much like Sacher-Masoch, Krafft-Ebing had a wide range of interests in his field, but it would be sex that gripped his attention, sex that cemented his reputation and notoriety, and sex that sold books. Lots and lots of books.

Our curious Krafft-Ebing had a full beard and pensive, sad eyes. He had been a gifted student and was a dedicated family man. Nothing so far suggested a particularly revolutionary streak in him. So you'd be forgiven for not guessing that our hirsute, wool-coat-wearing Austrian professor was about to write one of the most lurid, controversial works of his time, a work filled with new names for things that people had known about for ages but hadn't quite mustered up the courage to talk about in polite company.

The book was called *Psychopathia Sexualis*, a text chock-full of people describing their seemingly atypical sex lives. (An American scientist, sixty-some years later, would prove that these sex lives were a lot more typical than believed, but we'll get to him soon.) It was full of truly kinky shit—beatings, role play, fetishism, genital torture. Even in the rapidly expanding world of European sex studies, no one had compiled anything quite like it before. It would come to be regarded as the most important work on sexual pathologies of the age and for many years to come.

Publishers raced through twelve German language editions of *Psychopathia Sexualis* during Krafft-Ebing's lifetime. Translations into French and English were hot on the tail of the first edition. Other translations followed soon after. Although Krafft-Ebing's preface and analyses were all written in German, the salacious case studies were written in Latin to keep the juiciest bits out of the hands of the laypeople. It did not work. Anecdotal accounts from this period mention a conspicuous increase in sales of Latin-to-German and Latin-to-English dictionaries. People wanted their erotica. People would fucking learn Latin if they had to! The public pushback from Krafft-Ebing's more conservative peers and the German Purity League did little to curb sales. Sadly, we don't have access to the kind of detailed publishing data that's available today for books like *Fifty Shades of Grey*, but we do know that *Psychopathia Sexualis* sold like hotcakes, if hotcakes existed back then (they did). It sold as well as the steamy potboilers of the time. It sold like *Venus in Furs*.

Krafft-Ebing had certainly heard of *Venus in Furs* by now. He certainly knew about the former faculty member of the university in Graz where he worked who had made such a name for himself describing in tantalizing detail scenes of bondage, humiliation, and brutal floggings. Like any celebrity and shitty boyfriend, Sacher-Masoch had made plenty of enemies along the way. His disdain for convention, his flamboyance, his philo-Semitism, his string of jilted lovers—all left him open to attacks, and he was attacked regularly in the press, which usually just increased his book sales. But behind the scenes something else had been brewing. At least one anonymous source approached Krafft-Ebing to let him know that Sacher-Masoch's fiction wasn't all that fictional. Some historians have surmised that his wife, Aurora, the new "Wanda" in his life, spilled the beans. She definitely did not hold back any details in her *Confessions*, published just a few years after Sacher-Masoch's death. According to historian Harry Oosterhuis, it was an anonymous source from Berlin who first drew Krafft-Ebing's attention to the works of Sacher-Masoch and the underground culture of men drawn to pain and domination.

Regardless, Krafft-Ebing got the inside scoop from someone as he was compiling stories for his sexy opus, and the rest is history:

2. MASOCHISM. THE ASSOCIATION OF PASSIVELY ENDURED CRUELTY AND VIOLENCE WITH LUST.

Masochism is the opposite of sadism. While the latter is the desire to cause pain and use force, the former is the wish to suffer pain and be subjected to force. . . .

I feel justified in calling this sexual anomaly "Masochism," because the author Sacher-Masoch frequently made this perversion, which up to his time was quite unknown to the scientific world as such, the sub-stratum of his writings. . . .

During recent years facts have been advanced which prove that Sacher-Masoch was not only the poet of Masochism, but that he himself was afflicted with this anomaly. Although these proofs were communicated to me without restriction, I refrain from giving them to the public.

Imagine waking up and finding your name in a textbook. Not like a name that you share with lots of others, but your specific name, with reference to you specifically. Imagine that no one had interviewed you for this textbook, that no one had even given you a heads-up that you would be making a feature appearance in this textbook. Now imagine that, improbable as it might seem, people were reading this textbook in every park, bar, and coffeehouse in the entire fucking country. Leopold, proud inheritor of the noble family name Masoch, a name gifted by his beloved and now deceased mother, instantly became known as "that masochism guy in that freaky sex book." He was livid. It had been a rather open secret that he may have shared some interesting proclivities with his fictional protagonists, but being named for a sexual pathology was altogether different. The Victorian veil of propriety had been ripped right off. A secret informant had been involved. Someone at his old job was responsible. And what's worse, Sacher-Masoch's romantic ideal, which had inspired so many torrid affairs and had sold so many books, was now a condition, a mental illness. It had been named, marked by scrutinizing eyes, and so had he.

Unscrupulous as that particular move may have been, Krafft-Ebing's motives with *Psychopathia Sexualis* were not nefarious. He was a white male scientist from the 1800s, so naturally there is plenty of misogyny and horror of masturbation in his work. Make no mistake, he was a product of his time and place of privilege. Much of his writing was riddled with Victorian prejudices and bourgeois hypocrisy. But Krafft-Ebing also worked tirelessly to decriminalize mental illness and shift the focus of jurisprudence toward treatment and rehabilitation. Considering how hotly contested these ideas are still, Krafft-Ebing was on the cutting edge not only of the psychiatry and neuroscience of his day but also of criminal activism. He was likely influenced by his time at the University of Heidelberg, where his maternal grandfather, Dr. Mittermaier, served as the "German attorney for the damned," championing the legal rights of those condemned to ostracism by a still puritanical society. One such group were sexual deviates, who received harsh treatment in the courts, notes Klaf. Sexual practices of all kinds were prosecuted ruthlessly at this time. Homosexuality, transsexuality, masturbation, oral sex, you name it. If you weren't trying to knock someone up in the context of a state-and-church-sanctioned marriage, there was probably a law on the books about it. Called in as an expert witness in many high-profile sexual deviance trials, Krafft-Ebing set out to use science to advocate for rehabilitation rather than imprisonment.

In *Psychopathia Sexualis*, Krafft-Ebing let homosexuals, fetishists, and masochists describe their life stories in great detail, giving voice to voiceless sexual deviants. He did a lot of good in his lifetime, but that didn't offer much solace to Sacher-Masoch, the freaky sex guy. It didn't suit a wild romantic to be cataloged and categorized like this. And the patriarchy of his day never forgave him for submitting himself to women. Once his ex-wife's memoirs came out confirming his masochism, Sacher-Masoch's reputation was wrecked. As writer Ciaran Conliffe puts it so succinctly, "During his life he had been known as a great writer, but in death he became a punchline."

Ironically, it's safe to say that way more people know of Sacher-Masoch's works today than would have if he hadn't found his way into *Psychopathia Sexualis*. Many people learn about the eponym first and then discover his work. His pervy counterpart, the Marquis de Sade, at least enjoyed the privacy of the grave by the time Krafft-Ebing used his name to coin the term "sadism," also included in *Psychopathia Sexualis*. Both men had distinguished literary careers but were truly immortalized by Krafft-Ebing, who, just as ironically, almost no one knows about today. Maybe if someone had named a kink "ebingism," more people would look him up.

As time goes on, Sacher-Masoch's name continues to proliferate thanks to his eponymous "affliction," spreading outside academic circles and to a public hungry for sex and all the ways we talk about it. As I write this, actress and pop singer Sky Ferreira's eagerly awaited sophomore album nears its 2021 release. That album is called (shock of shocks!) *Masochism*. I gotta hand it to Krafft-Ebing, although naming a paraphilia after a living dude that you sorta know is ice-cold, in doing so, he unwittingly gave us a very useful word, the meaning of which has expanded over time, claiming new ground while still grasping firmly at its roots. Today, a masochist can be someone who works too much, a cuticle-biter, a ballerina, a grad student, an adventurous gourmand at a spicy salsa bar, a marathoner, and, yes, someone breathing shallowly through a hole in a shiny latex gimp suit.

It's interesting to me that there's a degree of cultural consensus around the word *masochism* and its modern usage that is so far removed from the specifics of its inception. Everyone knows what a masochist is: someone who does painful things on purpose. And yet, the word itself still carries so much of the taboo baggage that attended its birth. It's used to describe a broad phenomenon while the public remains squeamish about the shameful impli-cations of its origin. I think it's worth considering how that happened.

So I call up Dr. Stephen Stein, a historian at the University of Memphis and author of a forthcoming book on the history of the BDSM community in the United States, to ask him how public perception of masochism has changed since it was named as a pathology by Krafft-Ebing. On his faculty page, Dr. Stein is happily holding a big snake, which makes an enviable first impression that I both enjoy and respect. He tells me that he thinks this instance of naming is both helpful and harmful. "It's helpful because it does give people a language to finally discuss their desires," he says, refer-encing the rise of related magazine articles as periodicals became more com-mon. "Suddenly, the people into flagellation and corset training took over the correspondence columns." Stein is talking about the charmingly British phenomenon of the *Englishwoman's Domestic Magazine*, a seemingly proper publication that ran from 1852 to 1879 and that spent a curious amount of time describing domestic disciplinary actions. From an 1870 issue:

> I put out my hands, which she fastened together with a cord by the wrists.
> Then making me lie down across the foot of the bed, face downwards, she
> very quietly and deliberately, putting her left hand around my waist, gave me
> a shower of smart slaps with her open right hand. Raising the birch, I could
> hear it whiz in the air, and oh, how terrible it felt as it came down, and as its
> repeated strokes came swish, swish, swish on me.

While things on the Continent were getting saucy, the Brits certainly gave them a run for their money. England was awash in spanking tracts with titles both provocative yet formal, like *The Exhibition of Female Flagellants* and *The Romance of Chastisement*. Theresa Berkeley was a celebrated London dominatrix from the late 1800s whose house of pleasure was dedicated solely to flagellation of its patrons. The French even named this penchant for pad-dlings *le vice anglais*, or "the English vice."

Krafft-Ebing gave many people a formalized language to describe what they wanted. People could now know that they were not alone in the world. "On the other hand, then you get people being pathologized," says Stein. Which is not great. No one wants to be told they are sick for liking what they like (unless that's their kink!), and the stigma of mental illness has allowed outsiders to demonize people in BDSM communities.

In 1975, a notorious excoriation of sadism and masochism ran in the popular New York City alt-weekly newspaper, the *Village Voice*. It painted kinksters as fascists and murderers and asserted that masochists are all suicidal. The connection between pathologized sexual conditions and violent crime was not even subtle. Already on the margins of society, the S&M community was horrified with this kind of publicity. They pushed back. According to Stein, this is when the phrase "Safe, Sane, Consensual" emerged to identify a collective ethos. Organizers worked to ensure the phrase wasn't just hollow rhetoric. The National Leather Association led an initiative to root out domestic violence among community members. It was important for practicing kinksters and the public to distinguish between consensual play and abuse.

―――――――

Over and over again we see this trend of inquisitive academics casting light on masochism and human sexuality, followed by the inevitable public backlash. In 1893, the well-starched *British Medical Journal* said of *Psychopathia Sexualis*, "We have questioned whether it should have been translated into English at all." Sacher-Masoch's literary fame crumbled after being featured in Krafft-Ebing's opus, despite Herr Doktor's assertion that "the number of cases of undoubted masochism thus far observed is very large." But neither figure managed to inspire adoration and outrage quite like an American entomologist in the late 1940s. The world is indeed a curious place, and that a gall wasp researcher from Hoboken would find himself in this chapter is proof of it. The researcher was Alfred Kinsey, and in 1948 he didn't just upset the applecart, he set the applecart on fire and fucked on top of it while it burned.

At the time, Victorian social mores were being buffeted violently by social changes and new technologies, but the puritanical streak in the United States is as wide and deep as a mighty river. Son of a maniacally devout Methodist father, Kinsey felt this in his bones. He was born in 1894, just one year before Sacher-Masoch died. (I was shocked by this realization. They seemed to be so far apart in time, worlds apart. Sacher-Masoch wore cravats and rode horseback; Kinsey gave interviews on television.) Like Sacher-Masoch,

Kinsey was a gifted and hardworking student who refused his father's career guidance. Instead of pursuing engineering, he followed his heart into the wilds of biology, where he excelled. He brought a methodical rigor to his work that would earn him great respect, first at Bowdoin College and then at Harvard's Bussey Institute, where he took up his study of gall wasps.

Young Kinsey spent years collecting, measuring, and recording hundreds of thousands of specimens. All by hand, bent over his workstations. Hundreds of thousands. Imagine that. Krafft-Ebing would have been impressed. Both scientists contributed to their fields possibly even more through their methodology than the information they gathered. No self-respecting entomologist could spend that much time with a species and not think, extensively, about its mating practices. Kinsey and some colleagues at Indiana University, where he went on to teach, began to wonder why the same kind of scientific inquiries and methods were not being applied to the human species.

In 1938, Kinsey got his shot at tackling this issue when he was offered the job teaching a class on family and marriage. The class was formed in response to a petition made by the students themselves. The poor kids were drowning in a sea of ignorance about their bodies, sex, sexual health, and procreation. People at the time believed that masturbation caused impotence or, even more fancifully, pregnancy. There was no internet or Planned Parenthood for them to turn to. The curriculum of the day consisted of abstinence education and a book entitled *Ideal Marriage: Its Physiology and Technique*, which, needless to say, didn't offer much advice on having a zesty sex life. Kinsey's class, replete with detailed illustrations of copulation and frank discussions about contraception, became an instant hit. Hundreds more students enrolled.

Ever the consummate notetaker, Kinsey began conducting questionnaires of students' sex lives, and the project expanded rapidly with the financial backing of the Rockefeller Foundation and the National Research Council. After ten years of research, Kinsey and his assistants had collected over five thousand sexual histories. (Krafft-Ebing would have been green with envy. The twelfth edition of *Psychopathia Sexualis* had 238 case studies, considered an astounding collection for its time.) It was time to publish *Sexual Behavior in the Human Male*. Full of scientific jargon and clocking in at a whopping 804 pages, most people involved in the project expected modest sales to

other academics and collectors of novelty doorstops. But as I may have mentioned, people positively adore reading about sex.

The opus was renamed *Kinsey's Report* by a shocked and titillated public,
and it sold two hundred thousand copies in the first three months. (Sound
familiar?) Journalist Albert Deutsch compared the report's importance to
the works of Darwin, Freud, and Copernicus. It climbed best-seller lists and
inspired popular songs like Cole Porter's "Too Darn Hot" and a banned-
from-radio bombshell by Martha Raye called "Ooh, Dr. Kinsey!":

> *Now Kinsey met a man*
> *Who got his kicks*
> *Sticking his mother's sharp hatpins*
> *Into fine, little chicks!*
> *And when it was all over,*
> *This man would laugh with glee*
> *And say, "I did it to you, now you do it to me!"*

The report revealed all kinds of scandalous things to the American public.
A whole lot of people masturbate. A whole lot of people fool around before
and outside of marriage. A whole lot of people have queer sexual experiences. A whole lot of people enjoy pain. According to the survey, about half
the respondents enjoyed being bitten during sex. The list goes on. And on.
Kinsey's interview kit had over three hundred questions. Like I said, he was
a thorough guy. It being the late forties, with the Cold War looming and a
clean-cut, televised American identity on the rise, public backlash stirred
but did not come tumbling down on Kinsey's work just yet. The first edition of the *Diagnostic and Statistical Manual of Mental Disorders* (*DSM*) of
the American Psychiatric Association was published in 1952. It was like the
upright cousin to Krafft-Ebing's *Psychopathia Sexualis*. In its authoritative
pages were described many mental illnesses that Kinsey and company said
were common practice in the average American bedroom. It was a scandal,
an outrage! At least that's what moral conservatives decried. Sales continued
unimpeded.

Kinsey wasn't just a disinterested observer of sexual deviations. He was a
private practitioner and an outspoken advocate. In a 1935 lecture at Indiana

University, he laid bare the cause that would become his life's work. Discussing "perversions," he said, "The current use of the term, especially in regard to sex, for activities which are not approved by the mores of the day has no application in a scientific consideration of the problem." He beat this drum loudly throughout his career. This thinking marked a huge leap forward from Krafft-Ebing's work. As Kinsey's biographer James H. Jones notes, "While Krafft-Ebing's views were humane and enlightened for his day, they would not have offered Kinsey any peace of mind. Krafft-Ebing called for compassion and sympathy; he said nothing about tolerance or acceptance."

Homosexuality and masturbation were primary targets of psychiatry before and after Kinsey's day. In the house of his hellfire-preaching father, Kinsey himself had endured vivid onslaughts. Puberty found a sickly, often bedridden Kinsey in a turmoil of shame and natural passions. He desperately needed sexual relief and often fantasized about other boys. Tormented by conflicting drives, Kinsey began inserting straws and then larger objects into his urethra when he masturbated. Religious moralists had branded masturbation "self-pollution" since at least the seventeenth century, and in the nineteenth century, pseudoscientists took up their awful work. As Krafft-Ebing wrote in *Psychopathia Sexualis*, masturbation "despoils the unfolding bud of perfume and beauty, and leaves behind only the coarse, animal desire for sexual satisfaction." This had been a common trope for ages, inspiring NoFap crusades by everyone from ministers to dietitians and, in the case of graham cracker inventor Sylvester Graham, minister-dietitians. Kinsey sought to purge his guilt and sexual desires simultaneously with his masochism. The foundation of his crusade for sexual tolerance was inspired by his needless suffering, imposed on him by a puritanical father and society. His genital torture escalated over time, culminating in an unanesthetized self-circumcision in a bathtub with a pocket knife.

By the time the second half of Kinsey's research came out, the social terrain had become perilous. Given America's backward views on gender, *Sexuality in the Human Female* was destined to have an uphill battle. Now this pervy academic was claiming that people's daughters, sisters, wives, and mothers were masturbating and enjoying sadomasochism! Meanwhile, Joseph McCarthy and other high-profile conservatives had been fanning the flames of homophobia, correlating "deviant" sexual practices with (I bet you'll never guess) Communism. The Lavender Scare, as it later came to be known, ruined the careers of thousands of queer government employees when President Eisenhower signed Executive Order 10450, which barred homosexuals from serving in any government department or agency or working with any corporation that had a government contract. The year was 1953, the same year Kinsey's second report was published. Dr. Kinsey had also been weathering criticisms, some valid and some not, from the scientific community about his methodology and personal advocacy. Sales of the second report were underwhelming, and funding was soon withdrawn from his endeavors.

Under tremendous stress and suffering from lifelong health problems, Kinsey passed away in 1956. He did not live to see himself become a folk hero to numerous counter- and subcultural movements just a decade later. But no amount of bad press managed to bury his legacy. Like Krafft-Ebing and Sacher-Masoch, he threw a spotlight on the shaded recesses of people's sex lives and revealed that people get turned on by all sorts of things. His work helped normalize behaviors that, though ancient and nearly universal, often made easy targets in the Christian world. Kinsey named the unnamable and, in doing so, helped give a voice to the voiceless but also, unwittingly, gave fodder to their oppressors.

———

Something I try to notice regularly, both in my work and my personal life, is how attention changes perception. That the act of looking at a thing, naming it, describing it, and categorizing it can fundamentally change the experience of it. This is not to say that changes stemming from attention are bad or good or have any essential moral character at all. Attention can be paid with all denominations of emotional currency. It's just that giving attention is an action, and actions can cause change.

But.

The introductory context in which attention is first shone upon a thing influences how it will be received, both in the present and on future encounters. Since its inception, the word *masochist* has entered the public vernacular in a colloquial manner. People throw that word around to describe everything from workaholics to ballerinas to human pincushions. But, at the time of the word's conception in 1886, *masochist* described a specific sexual pathology. And this all happened during a time when discussions of sex and power were, well, complicated. For that matter, they still are. And I argue that the origin of the word still heavily influences modern Western sentiment about the propriety of pain on purpose.

Even though the word is in broad cultural usage and is employed to describe all manner of deliberately painful acts, when it comes down to it, the word *masochist* still conjures the image of a sex pervert.

From its origin as a pathology, to its current status in the *DSM* (which says, basically, that if it's not a problem, it's not a problem, but more on that in a second), *masochism* is a word that carries with it the imprint of the culture that created it, stigmatized it, and performed much hand-wringing around it. But it's true that we need a word to describe the pursuit of pain on purpose in any context, hence the retrofitting of a career-ending eponym from freaky sex guy to general-use pain pursuer. That we also still use it for the sexy side only compounds the confusion. But if you ask Krafft-Ebing, Alfred Kinsey, or E. L. James, you'll find that readers do not mind the word one bit.

There are vast similarities among getting spanked for pleasure, eating spicy food, jumping into cold lakes, and running a marathon. With this understanding, I propose that we do not let outdated, myopic associations between sex and shame impede our willingness to consider and explore the role of deliberate pain in all facets of life. Just because pain on purpose can lead some people to sexual arousal doesn't mean that all deliberate pain is inherently sexual in all contexts. In fact, although some of my consensual painful encounters have been explicitly sexual, many, many of them were not.

Nearly a century and a half after *masochism* entered the cultural lexicon, American views of sexuality still roil with guilt and shame. If you ask a gym buff who works out until their muscles are screaming whether they have anything in common with someone who likes to get the business end of a riding crop, chances are they will laugh uncomfortably and then change the topic.

Which is too bad, really. Overly conflating our knotted-up feelings about bodies and sex with our feelings about the pursuit of pain can cloud the waters of self-reflection. There are healthy ways to pursue pain on purpose, and there are unhealthy ways; being preoccupied with shame can make self-assessment difficult. For example, Christian and Anastasia of Fifty Shades fame explore all kinds of abusive partner tropes in *Fifty Shades of Grey*. Sacher-Masoch's own kinky pursuits were questionably consensual, depending on which sources you believe. Me? I took my athletic pursuits and cruel mastery of my body much too far, and it nearly killed me. Which leads me to ask: At what point does pain on purpose become a problem? Where is that diaphanous boundary between safe and unsafe? It must exist somewhere, a nebulous point where the sharpness of consequences begins to creep in. There is an enormous gulf between sensual pain play and dangerous self-harm, and to talk about the former without discussing the latter would be a great disservice to understanding pain on purpose. As we'll see in the next chapter, differentiating between the benign and the dangerous finds us wading into murky, turbid waters.

Chapter Six

WHEN THE LIGHTS GO DARK

BEFORE WE JUMP IN, I WANT TO INFORM YOU THAT THIS CHAPTER ASKS A few difficult questions that are difficult to answer. When is pain on purpose okay, and when is it harmful? And how can you tell the two apart? There is a difference between dabbling in pain that is ultimately benign and engaging in harm on purpose; that is unquestioningly true. And at both ends of that particular spectrum, the answers to those questions are clear as a bell. But in the middle? That's where things get more difficult to parse. That's what I'm going to explore in this chapter.

I wanted to go ahead and state that out in front. This chapter was hard to write; it's going to be hard to read, but honestly? I think that's okay. A life does not need to be sanitized before it's engaged with critically and thoughtfully.

With that in mind, here is what I want you to know in advance:

1. No one, at any point, is going to make the argument that self-harm "isn't so bad" because some people engage in benign forms of pain on purpose. That would be like saying alcoholism doesn't exist because

some people drink casually. Self-harm is dangerous, and this chapter is about how to identify when pain-seeking becomes a problem.

2. No one is going to make the case for benign masochism as a treatment for self-harm. Again, that would be like prescribing a single, daily nightcap to a compulsive alcoholic. Some people, myself included, have a history of catastrophic levels of self-harm. With that in mind, I have a deep curiosity around my current, hobby-level relationship with masochism, given that I used to engage with pain in a very harmful way. This chapter is an attempt to interrogate my own complicated and mutable relationship with pain and does not seek to imply a link between self-harm and benign masochism for all people. Lots of people come to masochism without prior pathologies! But some people do, and I'm curious about that particular slippery slope.

3. I was very, very sick, and now I am healthy. There's going to be frank discussions of eating disorders and self-harm in this chapter, and I want you to know that, despite the odds, I came out the other side of it just fine. I am, unreservedly, one of the lucky ones.

―――――

Okay, let's begin.

It's the back half of winter in Chicago, sometime in early 2007. I'm twenty-one years old, living in a third-floor apartment in Bucktown, in an old building with beautiful wood floors in need of repair and two roommates, both men from Craigslist who remain strangers to me. I recently spent a year or so playing Blue the Dog in the national tour of the stage production of Nickelodeon's *Blue's Clues Live!*, a role I landed accidentally after I went to the dancer call for the show. These days, I'm barely treading water in an expensive city, playing chicken with my death drive.

The days run together, but they all look something like this. I get up when it's still dark. My face is puffy, salivary glands jutting out from under my jawline like little walnuts tucked beneath my earlobes. I feel like shit. My heart sputtering around in my chest is a trapped chimney swift. I stand up too fast, and the tunnel closes in, sending me back to the mattress. I still have to go to work, though. Sluggishly, I try again, padding down the hallway to the small, shared bathroom. I splash water on my face and inspect the whites of my eyes for little red dots, the backs of my knuckles for scabbed-over scrapes. I run a flat iron over my temperamental bangs, but holding my arms up is tiresome, so I resign myself to getting fussed at for my lackluster appearance. Besides, there's usually a stylist at work who is willing to take pity on me and fix my hair before the boss logs into the security camera feed and phones us with her individual aesthetic critiques. She's a real treat.

I put on layers and layers and layers of clothes, but it won't be enough. I'm malnourished, sallow. I've stopped dancing for now; I'm too sick. I think I'll go back once I lose a little more weight and get my shit together, but what I don't know yet is that I never go back, not really. I'll do some shows here and there and join a modern dance company for a short spell, but the damage I am sustaining daily is too much for my body. It will nearly kill me. Soon, I will stop dancing completely.

I don't feel like drinking any water yet because I'm savoring the woozy feeling of dehydration, my parched body folding in on itself, stomach puckered. I am trying to remain as distant from myself as possible, to do everything I can to mute this feeling, but deep down I seethe. I am furious all the time, perfused with a benthic, corrosive rage. I keep my anger sealed behind layers and layers of impenetrable glass. It won't be until later that I will understand the abuse I suffered. All I know is that for now, I am living as a brain tethered to a body, looking desperately to escape. I need out, need solace, need comfort, need silence. I am sad, relentlessly lonely, and filled with hatred, entirely adrift from the life I'd built. Who am I if I am not in the studio for eight hours a day? What is pleasure if I am not dancing? What is left for me? There is a hole in me and I fill it. I fill it again and again and again and again and again, and each time, I scrape it empty once more with vengeful desperation.

I walk through the searing cold to my job as a desk girl for an upscale salon and spa. It's about 6:45 a.m., and I have about a mile to go. I am cold in a way that shuts my brain off, emptied of anything but hate and sorrow. I cry, and the tears freeze on my eyelashes. I resolve to leave this place. During the first half of my shift, I eat raw almonds and dried cranberries, set out in a bowl for the rich white ladies coming in for expensive foil highlights and performative fawning. I offer to pick up a colorist's lunch order from down the street because I know that particular sandwich shop has a single-occupancy bathroom, which is great for public barfing. I chug some cucumber water from the spa, remove my billowing smock, and trot down the street, happy to be close to release. I pop into the shop, which has a name like "Fat Tummy's" or "Thick Belly," which feels ironic to me as I slip into the bathroom. I can throw up pretty quietly and often without using my hand, but there's not much food in my stomach, so it takes some doing. My nighttime binges are different; I can pull the heft of them up with sheer muscle memory. I'm excited to do that later tonight, but now I'm annoyed at how hard I am retching to produce such a stupid, middling amount of uncooked trail mix. (Despite everything, I try not to be obvious about my incredibly obvious mental illness.)

The purging hurts, and I feel swallowed with disgust for myself, face perched above a public toilet, pathetic girl. I retch and retch until I taste bitter neon stomach acid and my knees buckle and I catch myself, poorly, the floor coming up to my face. It's cold, and my hearing turns to the sounds of ringing, and I sigh, emptied of anything but the ringing, ringing, ringing. That ringing, feeling feels so good to me, it is the only peace I know. These days, I'm doing everything in my power to chase it and it's killing me.

In the pantheon of my lifetime love affair with masochism, sown in me during years of ballet, born in me through history and chance, patterned for me and taught to me, embraced by me, loved by me, desperately cultivated and sought by me, this is the time where things are really bad.

I get up off the floor. My knuckle is bleeding, an amateur tell. I wash my face and rinse my mouth, reapplying my lipstick with a trembling hand. I pick up my coworker's to-go food, plucking a peppermint out of the courtesy tray and flirting easily with the bored cashier. I float back down the street to work, stopping once or twice to steady myself, but generally feeling the buzz of violent emesis, like someone had finally uncorked me and let the putrefied ooze of my starving brain seep out my ears. I work the rest of my shift hungry. The hunger hurts, but I feel like I deserve it. It also makes me feel accomplished, like I'm doing something right.

Around five thirty, I clock out, bundle up, and walk to my second job, which is at an ice cream parlor. Sometimes I'll eat a sandwich and throw it up before I head into Job Two, but it's so cold, and I just don't feel like calculating the logistics of selecting food, eating it fast, and finding a good place to barf in the middle of the Bucktown dinner rush. I clock in by six and keep my winter coat on, shivering as I scoop ice cream for people peeking through their scarves and winter hats. Time is as slow as it's ever been, and I am starving, irritable, smiling demurely at customers, singing stupid ice cream jingles every time someone tips me. The cops come in because they know the store owner says not to charge them. They order complicated ice cream treats and tease me and give me five dollars to "make a Beyoncé song about ice cream." They ask me what my favorite flavor is, how I stay so skinny working in an ice cream store, how old I am.

When they leave, I wait the remaining hour until close and lock the doors. It is only then that I start to eat, swallowing urgently as I run through the end-of-day tasks with listless efficiency. I get a shift ice cream and chew Oreos and roasted nuts and refrigerated cookie dough while I sweep the floors at midnight. I bleach the toilet that I'm about to throw up in, and then I bleach it again. We work alone in the winter, so I know that I can safely eat and barf in the employee bathroom without detection. I also know that lonely shifts mean that I can take a pint of ice cream to go without anyone knowing. (I'll point out that I did not truly believe my relative stealth was fooling people about the state of my mental health and ability to feed myself. It was more like I believed putting effort into at least attempting to conceal my eating disorder decreased the chances that someone could catch me red-handed and confront me about it. I'm sure people knew I was sick, but it's a lot harder to stage an intervention if there isn't concrete evidence. No body, no case.)

I stuff some ice cream into my bag and, roughly eighteen hours since I left my apartment that morning, I set the store's alarm, lock the doors, and head home. It's so cold on the walk back that I start crying again, which only brings the chill in deeper. I stop at a convenience store and buy bread, candy, diet soda. I shoplift some nearly expired sandwiches. It's an expensive, wasteful lifestyle, bingeing and purging all the time, and I pick up a dedicated side hobby of petty theft to fuel my addiction. I hate myself for it and am bundled in my shame. It doesn't deter me, though.

When I get home, I start in earnest. White bread with room-temperature margarine spread that's been sitting open on my dresser all day, ice cream stolen from work, stale turkey and cheese sandwiches (also stolen), fistfuls of chocolate, baby carrots and mustard, jarred queso with my fingers, cake frosting, plain tuna out of the can, diet hot chocolate mix (dry), bags of honey wheat pretzels, Pop-Tarts, Combos, crackers, bagels, bananas with peanut butter, uncooked flavored oatmeal packets, egg salad that is definitely off. I eat voraciously, wolfing down mouthfuls of silence, stopping again and again and again to barf into trash bags, to chug soda and barf again, to keep barfing as fast as I can eat so nothing gets digested, nothing gets to stay. On my knees, face aimed over an open plastic bag, sometimes the puke comes easy; sometimes I stick my hand so far down my throat that I feel cartilage, and I retch and heave as water from my eyes plops theatrically into the sea of barf that flows from me. Dry carbs congeal into glutenous lumps, and a poorly masticated plain bagel gets stuck in my throat. I keep trying to bring it back up, retching, feeling the pressure build in my body behind the bagel like a kinked fire hose. It's stuck in a weird place, making it hard to breathe, and I gasp, purple with exertion, desperate to draw this offender from my chest. It finally bulges, and the sensation of it coming up my esophagus is like a boulder being passed through a snake. I feel like my throat is tearing and tears stream down my face and the relief of it all nearly brings me peace. I'm so high from this violence against my body that I am swaying, even on the floor. I pull another dry bagel from the bag and keep going.

I know what everything tastes like on the way back up. At this point in Chicago, I have had an active eating disorder for a decade, since around fifth grade, though the tendrils of it stretch even further back in my memory. When purging, I play detective, making sure that all the food that goes in goes back out again, paying close attention to the strata of flavors in my puke. I can do 10,000, 15,000, 20,000 calories in a session, chewed and swallowed and regurgitated as fast as I can make all these things happen in a row. I'd probably have a good shot on the competitive eating circuit. I chew, blissful as an unfettered cow. I cry while I throw up. I laugh. I stumble, fading in and out of reality, electrolytes all fucked up, pancreas eating itself, nasal passages full of bile. I run out of food and throw up the last of it and chug water and throw up the water and chug more water and throw up more water and chug more water and throw it up and I'm crawling into bed and I'm hoping my alarm is set and I'm falling through the sheets and I get up once more to continue and I drink more water and I throw it up in great, splashing heaves and I do this until I pass out. It is the only peace I know.

It's very dangerous to throw up water in this way. It can disrupt the body's electrics to the point of cardiac arrest and death. The doctors later told me that I damaged my heart doing this.

Bingeing and purging provides a continual cycle of punishment and reward. The bingeing fills me with loathing but also, of course, pleasure. My body is in a long, slow spiral of starving to death. I am very underweight, well below my body's set point, which makes me panicky and obsessive about anything food-related. Eating feels good, chewing and swallowing and sugar and satisfaction and the bodily peace that comes with nourishment. Snacking, feasting, nibbling, gobbling, licking your fingers, washing it down. But this isn't eating, this is bingeing; at least, that's what I tell myself. The self-hate spirals through my head, *You disgusting pig you fat stupid girl you waste of human life people are starving and you are a glutton a monster a failure.* I am brimming with loathing, shame, fear. I eat until I am physically in pain, stomach so distended that I struggle to breathe. I think about a picture I saw once, the body of a dead girl, slumped over her toilet. Thin, but for a bulging middle, flesh already mottled with the signs of death. Her stomach split inside her when she was purging and it killed her. I think about her and cry and I cry and I hope bitterly that if this kills me, I'll die after I've already thrown up. I think about her constantly, pummeled with grief for this stranger. I push the grief away and fixate coldly on how my corpse might look when my actions reach their obvious conclusion, as if this is really at all about my appearance. (It's not.) I think about the girl all the time. She looks like me. She breaks my heart. I desperately don't want to die, and still I court death with reckless urgency. I feel rotted out from the inside.

I'll do this all again tomorrow, and the next day, and the next day. On days off from work, I do nothing except hurt myself. Eat food, throw it up, and pass out, body singing the cheater's euphoria of this grim release. As many times as it takes. I take long, meandering walks, stealing and buying food, throwing it up in various bathrooms, entire neighborhoods mapped out in my head based on places I can binge, and places I can purge. It is painful, this pushing up against the limits of my physical self, and I am high from it. I can do it for hours. It is what everything else in my life revolves around. It almost kills me several times over. And if this wasn't enough, because surely it wasn't, most days I also cut myself vigorously. Years later, my body continues to bear pale reminders, silvery seams from where I used to dig around inside my body with nail clippers.

Why? Why do I do this? Yes, anger. Yes, hatred and a desire to harm myself. But I believe that this was also a desperate attempt to medicate myself. To blur the enormity of the hurt I carried with me, hurt that I am still quietly and meticulously working through with mental health professionals to this day. I think I desperately wanted to feel something *else*. To feel bad to feel better, to feel something to feel nothing.

Or maybe it's 2005, and I'm on tour with *Blue's Clues*, starving all day and forcing myself to consume and vomit half-eaten food off of the dirty dishes on the room service trays that sit outside hotel doors, creeping through indistinguishable hallways in the late hours of the night. Maybe it's 2001, and I'm at ballet boarding school, throwing up my dinner before rehearsal, throwing up my breakfast before school, throwing up my lunch, throwing up in the middle of the night, throwing up in the shower, throwing up in a gas station, throwing up in my boyfriend's house, throwing up at home on Christmas, throwing up in a closet, throwing up in my hands, throwing up throwing up throwing up throwing up. Maybe it's 1996, and I'm ten and discovering that not eating makes my stomach feel like knives, but it's worth it to me, because suddenly I also feel powerful. Yes, it was bad then too. But nothing like it is in 2007, in that Chicago apartment. I harbored a long and deeply held eating disorder for most of my life at that point, but dance had always kept the real darkness slightly at bay. Without my sense of self, without the ridiculous discipline of ballet or the purpose it offered me, I was suddenly more empty and more angry than I'd ever been in my whole life. My interiority was completely obfuscated by a long, sustained scream, and pain offered me a reprieve from it. I used the instrument of my body as a muffler to my pain. I conspired against myself.

That is no longer my life. Therapy, community, and one thousand buckets of white-hot swaddling love tether me to this earth and teach me every day how to walk it. I am no longer consumed with my need for oblivion, no longer planning each moment of my life around my compulsion for suffering. And yet.

And yet!

I still seek pain. I structure small corners of my life around it. I get off on it, I play with it, I built my fucking career on it like a concrete foundation. Feeling bad and then better is still a game I play, a crutch I use, a treat for myself. What, then, is so different about my life *then* versus my life now?

My experiences with harming myself are not unique. "I feel like you and I have had a very similar experience with self-harm," my college lab partner Anna Gioseffi tells me over email.* It's important to note that Anna is a wild and brilliant force, one of those people who is great at too many things, and if she just had the courtesy to be an asshole, you could hate her for it, but she is wonderful, so you are doomed to adore her forever. Back in the day, we sexed a lot of fruit flies and did a lot of whip-its, and I have canonized her in the pantheon of my loved ones for making a fanny pack with a clear viewing window so she could show her pet rat the world. She is singular. Like mine, her relationship to pain is complicated, spanning the destructive to the play-ful. As she says it, "I think the line between harmful and benign masochism is very blurry!" It's a challenging thing to explain to therapists. "Their usual advice is 'keep an eye on it, and try not to do it,' which is less than helpful." Surely there is more specific advice to be had than that?

Anna started self-harming the summer she turned twenty-one. With a history of mental illness and a failed pharmaceutical intervention, she took a summer internship in Albuquerque, "away from all my friends and family, living alone in an apartment." From there, she says things just kind of boiled over. "I started cutting my forearms. I'm a creature of habit and pattern, so I had a sort of ritual for it." For her, it was about facilitating an emotional release. She said she cried a lot and spent time lost in the aesthetics of it. It was a time of deep depression and despair. Her cutting quickly spiraled, be-coming unavoidable.

Back in Florida, she kept up her new habit but had to be more discreet around her roommates. Later that semester, she dropped out and moved in with her parents, who tried to passively control Anna's self-harm with in-creased and invasive surveillance. That didn't work, but soon a new relation-ship gave Anna an outlet for emotional processing and accountability, which

---

* Anna uses both she/her and they/them pronouns; for clarity, we are just using one set for the interview.

ultimately helped her stop cutting. "It was kind of the pressure I needed to break the habit. It definitely felt like an addiction in many ways. Felt like something I could do in dark moments to make everything stop."

One of the things that makes Anna's story so interesting to me, friendship aside, is the way it parallels a lot of my own pathological history with pain, as well as my playful present. After a harrowing experience with self-harm, Anna has also moved on to different types of pain on purpose. "I have a close relationship with pain still, though the tone of that relationship has shifted." Anna, or rather, Anna Phylaxis, as she is known by her audiences, is a drag performer with a penchant for sideshow stunts. "I regularly use physical harm in my life, both on and off the stage."

She says her dalliances with pain offstage these days are very tame. "I've done some body modification and gotten tattoos and piercings over the years. I'm also a fan of BDSM," she adds, expressing a preference for marks like bruises and rope burns. "Overall, I think these things in my 'real life' are very benign. There's no risk of (or desire for) actual or lasting harm, and the intention is really focused on pleasure in one way or another."

Performing, well, that's another story. "Onstage, things become a bit more overtly harmful, though the intention is still not actual harm, and I consider safety before I do anything." Her stage arsenal includes temporary piercings with blood drips, a medical stapler, a hot glue gun, and her latest feat: blockhead. "I can now hammer the nail in!"

(*Blockhead* is a classic sideshow stunt wherein the performer puts a nail all the way up their nose.)

For Anna, performance-based versus private masochism fulfill two very different needs. "Offstage, it's entirely for myself. Fulfilling my desire for pain as pleasure helps me feel connected with my body, which is something I've struggled with most of my life. Onstage, it's definitely an emotional release. I'd say in that way it's more closely related to the patterns and feelings of the self-harm I used to engage in, though my actual intent is quite different." Drag is a release for Anna, and she describes it as an opportunity for her to vent anger and frustrations. "Adding in the self-harm element is just enhancing that. The audience response to my performances is an important part of that emotional release. Without their energy and shock, it doesn't feel like as much of a relief for me!"

So, what's the difference? What changed between the obsession with cutting and self-harm and the pain in her life today?

"I think many people (maybe even most people) engage in intentional pain in one way or another in their lives. For some people it's cutting, for others it's BDSM, or in many ways drinking or using drugs to excess is self-harm too. Pain seems to have this taboo joy that reminds us that we're simultaneously fragile and strong to be alive. I think what really determines if the process of self-harm is unhealthy versus innocuous is the intention and level of control we have over it."

But where, exactly, is that line? To find out, I spoke with Dr. Hane Maung, a philosopher of medicine and former psychiatrist, about the distinction between unhealthy and innocuous. Obviously, in certain cases, it is very clear which end of the spectrum a given action inhabits. Hot sauce? Fine. Catastrophic bulimia? Not fine. But where is the defining criteria about the murky middle? All of this feels too close together at times, too dangerous. How do we know what is okay?

"What is it that makes it specifically a mental or psychiatric problem, as opposed to other kinds of problems in living, or even just kinds of nonproblems . . . part of normal human diversity?" Dr. Maung muses, repeating my question thoughtfully. He says that, historically, traditional theorists have suggested that whether or not something is a disorder (or a problem) could be answered precisely and scientifically. A definite line in the sand kind of situation. However, he says, more recent philosophers are arguing against that notion. "Instead, we contend that whether or not something is a disorder also depends on value judgments, such as the judgment about whether the condition is associated with unwanted suffering; whether it interferes with a person's goals and desires," and ideas on how best to address such problems. Basically, is it causing suffering? Is it harming you? What role does the action play in your life, and how do you feel about it?

Like Anna, Dan Block is a sideshow performer, or in layman's terms, "I hurt myself in an entertaining way for money." He's industry famous for his rare take on the blockhead act: he hammers a nail into his face and then electrifies it with a stun gun. Dan knows I am a gorehound and answers all of my questions about how he can shove an endoscopic camera through his face (answer: he gouged his nasal passageways open with a progression of larger and larger nails), what the trick is to snapping a mousetrap on your tongue (there isn't one, but you shouldn't do it more than three times an hour or your tongue will become grotesque), and how to put out a butane torch with your tongue (for liability reasons, I'm not going to tell y'all how to do that).

Like Anna's, Dan's relationship with pain has changed from destructive and dangerous to something more benign. Whereas before he used his side-show stunts to legitimize and obfuscate the nature of his self-harm, being reckless with his body under the guise of *work*, these days, it doesn't feel like destruction anymore. He's not hurting himself because he's trying to be numb, the way he used to. So, what changed?

Therapy, he said. That, and getting a dog.

What does he get out of pain now? "An endorphin release!" He laughs. "Now, the pain compared to the joy and wonder [of the audience] usually is overshadowed."

Dan and Anna bring up questions for me because even though the ways they've engaged with pain have changed, they still engage with it. Before, the self-harm, be it public or private, carried with it a dangerous weight, an obsession, a necessity, but now they perform feats of pain that are markedly absent the despair and desperation of previous painful acts. To me it sounds like a slippery slope.

But, honestly, isn't it? That some people can take the act of masochism to harmful depths, only to pull back and develop a manageable, pleasant relationship with pain is not evidence that everyone can make that switch. For many people, I imagine, that would not be a safe outcome. It is merely curious to me how many of the people I have spoken to in the course of writing this book have done that exact thing, myself included: perching in their own little nook on that slippery slope. For me, it's as if the deliberate infliction of pain is not the sole root of the problem, but rather the context of the act, and the emotional resonance it carries, determines the relative health and acceptability of each specific instance of pain on purpose. Something about the difference in seeking harm versus seeking pain.

"I was never disturbed or embarrassed by my self-injury," writes Darien Crossley, a visual artist and dear friend of mine. Darien has dashing good looks and an earnest, cherubic charm. "I did understand that it was so-cially unacceptable. As long as I was able to, I kept the evidence of it secret, not because I was ashamed but because it belonged to me." Later, after be-ing admitted to inpatient treatment for their eating disorder, their coun-selors, as something of an afterthought, acknowledged their self-harm as problematic.

Darien's relationship with self-injury began when they were thirteen and followed them into early adulthood. "I was an anxious kid. Definitely a big worrier, very sensitive, very prone to perfectionism and magical thinking. My intrusive thoughts were distressing, but I didn't have the language to reach out for support when I was younger. I didn't sense that the adults in my life would be able to support me if I had known how to talk to them about it." They describe a sort of internal chaos that was cumbersome, an impediment. "There was a lot of movement and energy inside my mind, always, and I was trying very hard to keep things running smoothly for everyone."

Darien's cutting was similarly born of internal frenzy. "It was dark! My home life was chaotic and puberty was (so!) chaotic, and the first time I cut myself I was just curious to see whether it was possible. It really goes against every instinct that we have towards self-preservation, to cut yourself. It's a profound act of violence towards the self. And although self-injury was sort of vaguely in the cultural consciousness in the mid-2000s, there was not a lot of information about it at the time, so I am thirteen and experimenting with control and my body and frankly I felt like I had discovered a super-power. Knowing that I could cut myself made me feel strong: my secret talent was the ability to defy biology and withstand pain."

Lack of supervision at Darien's school gave them the privacy to experi-ment with and cultivate the self-harm behaviors that they began using to manage their anxiety. "Cutting myself allowed me to channel my worries into something physical and manageable. It felt very grounding to me. I was able to focus on the sensation of pain and nothing else. Sort of a tran-scendent, out-of-body experience. (See: endorphins.) I also thought that the whole thing was fascinating: the way it looked and felt. The way I healed. It was evidence that I was made of meat and stuff—I thought that was surreal and fascinating."

When I was cutting, I was similarly transfixed by the meat of my body, how strange that I had this corporeal form to tend to, that not only was I connected to this meat sack, I *was* the meat sack. My meat sack? Couldn't be true. And yet.

Darien said that on a practical level, self-harm fulfilled an emotional role. "When I hurt myself, I felt safe. It alleviated the 'doomed' anxiety feelings. It also felt like good karma, if that makes sense? (Magical thinking.) I hoped that by causing myself pain, I could prevent future pain. By hurting myself, I hoped that whatever order existed in the universe would determine that I had paid my dues and spare me." A sentiment we've seen before in penitential flagellants and hungry nuns.

For the next ten years, Darien intermittently engaged with self-harm. "Over time I did it less regularly, but at the same time my relationship with food and exercise became more damaging." By fourteen, cutting happened weekly, but by their late teens, their eating disorder had taken over completely. "Starvation was the tool that I used to control my depression and anxiety. Ironically, my food goals and rituals felt completely out of my control and caused me much more anxiety—the rules were always changing, the relief came less and less frequently."

Today, Darien's life looks different. Several years into recovery from both their cutting and their eating disorder, they no longer have the impulse to cut themselves. "I have more support in my life; I am still very anxious, but I can handle that better than I could before. I have found medication which helps with my anxiety. I think the medication was a big step towards moving away from self-injury. I have friends and I smoke weed."

Darien's relationship with pain continues, but they have found what they describe as more "socially acceptable ways to engage with that need," such as art ("just hours and hours of tiny lines"), video games ("I'm fucking bad at them and I love losing and losing and losing until I win"), and BDSM, which they say serves similar therapeutic purposes for them as self-harm. "I fall pretty solidly into the submissive/masochistic category, sexually, go figure. It's funny to think that, as someone obsessed with avoiding pain by controlling my mind and body, I feel most comfortable allowing others to control my body and cause me pain. Building new associations, baby, new neural pathways! The things that I want sexually ebb and flow, but I've noticed that my need becomes more explicit when I am feeling bored, unexceptional, or worried."

So, what's the difference for Darien between pain *then*—chaotic, dangerous, consuming—and pain *now*? "The only difference between my maladaptive self-injury and my sexy pain on purpose is, I feel, the intention behind it." Referencing a painful, pleasurable encounter, they said that "engaging with my partner was the same pain on purpose, but it was coming from a place of tenderness and connection and compassion towards myself."

I sometimes have reservations about sharing such delicate and potentially triggering personal stories with the public, especially because there are no easy answers. Within an addiction to alcohol or other drugs, there is an option to go cold turkey to stop forever. With compulsive behaviors around food or bodily sensation, recovery requires learning how to manage the middle ground and learn healthy habits of existence. I had a near-lethal obsession with throwing up, and yet, if I was to survive, I had to learn how to eat food in a healthy way. (It's taken over a decade, and I will probably always be working on it.) I have also, along with so many others, found a safe middle ground for myself when it comes to my high-sensation-seeking behaviors. I recognize that for many, as with drinking, there is no safe middle ground at the extremes of sensation; the only place for them to be is as far away as possible. I support everyone who makes such decisions. My desire to understand my own confusing feelings about pain-seeking is in no way an attempt to make a case for moderation in all things! Pathology is thorny, life is messy, and abstaining saves lives.

I'm just curious about how people like me manage the project of living in a body. For Anna, it comes down to therapy and paying attention. "For me, that mostly means being aware of my emotional state when I engage in these activities," she says, referencing pain for fun. "While it's usually fine, I wouldn't want to do it when I'm feeling depressed or stressed for fear of awakening the addict."

I know that this slippery slope is always there, in its way. I don't think that self-harm exists *entirely* separately from more benign and less harmful types of masochism. I think, as with substance abuse and pathological gambling, there's a continuum. Anecdotally, it seems like we all have our own levels of interest, compulsion, and pathology. It seems that the way we feel while engaging with pain is helpful in determining whether or not the act is considered harmful. And, much like other addictions, "safe" pain on purpose is in no way a treatment for self-harm addiction, just like no one would say that moderate drinking is curative for all alcohol use disorders.

All of this is to say that pain on purpose can be done in healthy ways and unhealthy ways, and it can be tricky to spot the difference.

A major component of understanding and treating behaviors such as compulsive cutting is understanding what utility the harmful act is serving. Many of my less successful therapeutic experiences were marked by my treatment provider's inability to demonstrate sufficient curiosity about the motivations of my interior state. Which is to say, they assumed they knew *why* I was starving and cutting and *what* I was getting out of it and that their externally applied framework for healing was the only true and correct way out. Certain behaviors were Bad and represented a Problem, and providers had little interest in what purpose those harmful acts were serving for me. Their assumptions made me feel unheard and unseen.

In the 2015 paper "Non-suicidal Reasons for Self-Harm," researchers reviewed firsthand accounts that offered reasons for self-harm, noting that "one of the barriers to healthcare is a lack of clear understanding of the functions self-harm may serve for the individual." They found that despite the fact that the most widely researched reasons for self-harm were distress and social influence, many people related their self-injury to self-validation and a sense of personal mastery. Which indicates that many people may experience positive or adaptive functions of self-injury, beyond social effects.

If a patient self-reports benefits and no significant damage or negative impact on their life, how do you know whether their masochistic engagements are okay? Is a person running in a credentialed ultramarathon until they pass out okay? What about erotic blood play? Eating hot peppers until you sob and throw up? Biting off all your cuticles? Getting tattoos when you're sad? How do you distinguish between what is acceptable and what is not? How do you know when pain on purpose is bad, or okay? I talked to so many people who have transitioned from what they considered to be dangerous levels of self-harm to a more peaceful existence that still includes pain. I am one of these people! But I can't put my finger on the specific point at which "pathological" becomes "permissible." And it feels unsettling not to have a clear barrier.

I don't have an easy answer for this, nor am I trying to force some Grand Unifying Theory of Masochism. I'm just trying to understand my brain a little better. So I reached out to Dr. Allan House, one of the authors of the paper, to ask him what the difference is between self-harm and other types of more benign masochistic engagement. Where is the line between harmful pathology and a painful hobby, and how do you differentiate between the two?

"One of the other good/bad, detrimental/unproblematic markers is the degree to which any behaviour is compulsive or uncontrollable," Dr. House writes to me over email. "So, as people move from an active, engaged, enjoyed relationship with food, drink, sex, exercise, or whatever into one that's driven, exclusive, unmanageable, then we tend to see it as abnormal in some way and reflecting personal troubles."

Dr. House, who is Emeritus Professor of Liaison Psychiatry at the University of Leeds School of Medicine, specializes in understanding self-harm and suicide.

"I don't think there's a single defining feature [of self-harm], and these things overlap in the sense that some people do more than one," like self-harm and also extreme sports. "So the question is something like—what gets a particular action assigned to one or other of these categories?"

Dr. House lays out some pointers for parsing the difference between self-harm that is dangerous and pain on purpose that is not. "Self-harm is more about the harm than the pain—substitution with painful but nondamaging activities (squeezing ice cubes, ice-cold showers, and so on) rarely seems to work." That, to me, is one of the most clarifying explanations of the difference between innocuous and dangerous: Is it about pain, or is it about *harm*? The motivation of the person is important too. "The driver is personal and related primarily to distress, low self-worth, and so on. Excitement, arousal, sense of mastery, and so on can come but later and don't assuage the negativity."

He also points out that many of the painful acts that I cover in this book—ultra running, hot peppers, polar plunges, body modification, ballet, sideshow—are not as taboo as self-harm. "The activities you're interested in are, broadly speaking, socially sanctioned ones, whereas self-harm is seen as in some way outré and pathological—that's why it is often not disclosed."

Dr. House is quick to reframe my question of where the proverbial line is regarding deliberate pain. "Why is self-harm not seen as socially understandable in the way that these other actions are? Perhaps mainly because of its strong link to suicide, with various forms of mental disorder, and with other damaging actions like eating disorders and unhealthy drinking." Ultimately, it comes down to how the individual feels about their actions. "Self-appraisal is really important here—when people do seek help, they see themselves as having a personal problem/being troubled."

So, how often do people with active, obsessive compulsions to self-harm move on to more benign forms of masochism? Dr. House says he hasn't seen much of it in his career, but that observation comes with a caveat. "My personal experience is that the sort of active substitution you are interested in isn't that common. As people step back from self-harm, they tend also to step back from emotional use of alcohol and food; but then psychiatrists don't often see people change in these ways." That is, change is hard, and sometimes people can't do it.

Dr. House's work focuses on this ambiguous area between "good" and "bad" pain on purpose. His research is about understanding what people get out of self-harm and how they can achieve the same outcomes through less damaging methods. By considering the utility of harming behaviors, it reframes the problem to be solved. Regarding self-harm as a choice and an activity to be understood, rather than just a symptom to be jotted in a notebook, allows for more insight into the needs of the people who do it.

There is no clear demarcation between pain on purpose that is harmful and pain on purpose that isn't. There is no question that self-injury and deliberate pain can be detrimental to the self. People can and do suffer greatly at their own hands. But the existence of a pathological variant of pain on purpose does not encompass the diversity of all the other ways that people choose to engage in pain.

The actions and emotions involved in deliberate pain are not classifiable into a binary "good" or "bad" category, and you cannot necessarily tell which is which just by watching. One person's meticulously considered and enjoyed human pincushion act is another person's public engagement in despair-fueled self-harm, and who could tell the difference from the audience's vantage? It's not that it is never obvious what the situation is. Sometimes things are scary and harmful and feel bad to the practitioner, and the harm caused is unambiguous. But there is just so much gray area outside of that.

So, what changed for me personally? Therapy. Oh my god, so much therapy. I still like what I like. I run and work out hard. I have rough sex and find pleasure and fulfillment in certain painful activities that are thrilling and high sensation but ultimately not harmful. There is a conclusion to be made that working to heal my trauma and repair my relationship with myself is separating my desire for pain away from my desire for harm.

Darien echoes a similar sentiment. "My past self-harm was inspired by incredible fear. It ultimately prevented me from reaching out for help. Now, I incorporate pain into my life but don't rely on it to keep me safe or demonstrate my worth. I know that there is dangerous pain on purpose, and I think that what determines whether or not it is a problem is whether or not it causes problems. I am comfortable now in the fact that I still want pain, and I am comfortable with the ways that I find it."

For me, it is not a perfect balance. I still toe the line and sometimes take it too far. But I've shuttered the furnace of loathing that once powered my compulsions. My objective is no longer to rub against death, or suffer to the point of punitive disassociation. What I feel now when I engage with pain on purpose is not so heavily colored by the hatred I felt for myself then. It feels more like play. An adventure, a test, a game, catharsis, rapture: it feels like so many things, where once it was a singular force motivating the dissolution of my entire existence. It is no longer all-encompassing. I still get high from all of it.

I do yearn for oblivion, but, these days, just a taste. There is more for me outside the void than in it.

Chapter Seven

SOCIAL CREATURES

It's New Year's Eve, and I'm at a Russian bathhouse in Brooklyn with Rain.

Rain, who is very funny and very beautiful and very tolerant of my stupid ideas, has agreed to run into the ocean with me tomorrow for the Coney Island Polar Bear Plunge. Tomorrow we will greet the year 2020 in the salty ice water that laps the beach that stretches out in the shadows of sleeping roller coasters, and we'll dash into the sea with a throng of like-minded idiots. But today, today we suffer in a different way. Today, we favor fire. Tomorrow, well, that's for the other option.

The inside of the *banya* is brightly lit, with white plastic tables surrounding an empty cold pool. Televisions dot the wall, and groups of men sit together over beer and pelmeni, watching sports or Fox News or Rain in her bathing suit. At the far end of the room is an ominously sweaty door: the sauna.

It's very cold outside, and though it is warm inside the banya, I am eager to finish thawing out in the heat. I don't speak any Russian and feel fumbling and shy in an unknown place of yet known rituals. I open the door and wade into air so hot it feels like it is grabbing hold of my body, like it's giving me the Heimlich.

I don't know how to state this more effectively or more dramatically: it is the hottest room I have ever been in, hotter than any weather I've been exposed to, hotter than I would ever, ever think was safe. But I go inside anyway. Because there are people sitting in the heat by the coals, placid, red-faced, wearing little white hats. Rain and I climb up to a high bench and begin our perch in the inferno. I vastly underestimated how hot this was going to be. I am used to the piddling steam rooms at my local YMCA: gentle, warm, filled with little old ladies sitting on beach towels. *Ahh, I see. Now this is a sauna.*

I immediately feel like I can't breathe right. The air is so hot that it makes me cough, and I suppress the urge, taking shallow breaths and willing my chest-tightening response to stillness. No one else in here is coughing. My skin scrambles to start sweating, blood rushing to the surface. My brain feels like one of those egg timers that floats in the boiling water, slowly changing color and darkening as the egg cooks. The words *protein denaturation* flash in my mind. A sustained internal temperature of 108 degrees can cause brain damage in adults by unraveling the intricately folded shapes of our noggin proteins. I feel certain I am going to reach this grim benchmark in about thirty seconds. But I take another shallow breath to calm myself. None of the regulars are panicking. Everything should be fine, right? Slowly, my initial terror gives way to a grim, sweat-slicked resolve to stay. To enjoy it.

But I am **not** enjoying it. I will enjoy what comes later: pulling the rope hanging from the ceiling that dumps ice water all over my bright red body; laps in the cool water pool that pilfer heat from me like a welcome thief; the steam room so wet and clouded that I can only see inches in front of my face; the unbelievable sense of floating, gentle stupidity that overtakes my brain after several rotations of sauna–cold water–steam room–rest. It is a testament to the oppressive heat of the sauna that when I leave it, for the first time in my life, I find true joy in ice-cold water. Generally, I hate the cold, hate cold water, am a furious reptile of a person. But this heat is different. This heat overcomes my resistance to cold. This heat makes me long for it.

I struggle to be still in the sauna, to stay. Rain sits next to me, silent. I do not need to look at her to know that she is making a stern, stoic face, and I don't want her to look at me because I am clearly turning red and suffering. My body urgently wants to flee. My alarm bells are ringing, but there are others around me who are indulging in this ritual, and I decide to trust the process and ride it out.

It does smell divine. The crisp, vegetal aroma of dried oak branches used for *platza* feels cleansing and invigorating. (*Platza* involves getting whipped in the sauna with these bunched branches, but it is late on New Year's Eve, and they are not offering the service tonight. A shame. I think a little sharp pain on my back would help me endure this heat.)

I try counting in silence, holding my head in my hands, bargaining with myself. My pride dictates that I spend more than fifteen freaked-out seconds in the heat, so I settle into a few minutes' worth of dilation and dripping. Like all humans, I am taking cues from those around me.

There is a flurry of Russian, which someone kindly translates for me as a request to make the room hotter. Is that okay with us? Of course, I say. Naturally! What could be more agreeable than adjusting the dial from "Volcano" to "Surface of the Sun"? I must have been grimacing. Cover your head, the translating man instructs me, a private aside to a foolish tourist. I'd noticed some people were wearing hats, but did not understand them. The hats, it turns out, trap cooler air near the head. I twist a thin white towel around my hair and return to my focused breathing, bringing the heat as deep into my lungs as I can.

I feel like I am getting on top of this experience. A twinkling nugget of ego delights in my burgeoning mastery over my body's flee response. But my ephemeral moment of pride is rudely interrupted when someone brings a real live baby into the sauna. Toddler, maybe? I'm guessing eighteen months old, its little diaper bunching and twisting as it wiggles and tries to grab the beaded pink slides I bought on the street that morning when I realized I forgot my flip-flops. Someone gently swats the baby's feet with a piece of discarded dried oak branch, and the baby babbles and coos and fusses only slightly. It's clearly done this before.

I am being shamed by a baby at the Russian bathhouse. Once more, I adjust my expectations of what I can endure in this room. I make it a couple more minutes, my determination invigorated by the presence of a tiny child, then rush, desperate and faded, to the small alcove where I am forcefully drenched with the icy water. After some languid laps alone with Rain in the cold pool and a quiet soak in the steam room, I get back in the sauna. I want to say it gets easier, but that's not true. I just get better at knowing what I can handle. My heat tolerance, buoyed by groups of scowling men and also a baby, is increasing.

We do several circuits before showering and stumbling dumbly back into the cold. I sit on the curb and stare at my phone like a dog, fishing around my hot soup brain for the ability to order a Lyft and respond to text messages. I feel great, wrung out, buzzing, stupid. It's the last day of 2019 and sweating buckets of salty electrolytes has me feeling cleansed and empty, ready for the wide-open promise of 2020. But, as good as I feel now, I know that I'd never have gotten here if it was just me in that hellish inferno. Left to my own devices, would I have been able to stay even half as long in the heat of the sauna? I assure you with great confidence: absolutely not.

Like so many social animals, humans are constantly looking to each other for behavioral cues and social signals. It's a very normal, common thing: when humans are in public, we tend to modify our behavior to match the actions of those around us. We go with the group. (There is ease to be found in outsourcing decision-making, you know.)

For instance, say that you are choking. The fewer people there are around you when that ill-fated hot dog lodges in your throat, the higher the chance that someone will act to save *quickly*. Were you to experience the same emergency in a room with more people, the odds increase that it will take more time for someone to decide to help. That's right, the more people around you, the less likely people are to take fast action. Known as the *bystander effect*, it explains a kind of behavioral inertia in play; the more passive watchers there are, the greater likelihood that everyone will just keep standing there. This is important to remember so that we can overcome it. If you see a person being assaulted in broad daylight, you cannot assume someone else will act, especially if there is a large crowd nearby. It is incumbent on each of us as individuals to initiate action, and once someone starts to intervene, it is likely that other bystanders will spring to action too. We take our social cues from other people, but someone has to break rank and act.

Syncing with the group isn't usually as nefarious as ignoring someone in need of help. Often it's as simple as choosing when to cross the street or how fast to walk on a crowded sidewalk or how to select a seat on the train. As kids, we learn by patterning the actions and behaviors of our caretakers; as we grow, our ability to read and respond to social cues develops greater levels of sophistication and nuance so that we can function in society with relative smoothness and ease.

It is normal and prevalent that humans sync their bodies with each other, and it begins at the very beginning of life. From the start, infants and their caretakers develop a pattern of matching behavior, emotional states, and biological rhythms, known as synchrony, going so far as to sync their heart rates when playing, their bodies making tangible the bond that grows between them.

Face-to-face, people are prone to spontaneously coordinating their movements to match each other's, a phenomenon known as *behavioral synchrony*, and it happens both with people we know and people we do not. Matching footfalls when walking with a friend, syncing up brain waves when watching a movie with others, line dancing with a group of drunk strangers—all ways that human physiology and behavior facilitate connection and a sense of belonging. In fact, a 2018 study published in the *Journal of Social and Personal Relationships* found that simply imagining walking in rhythm with a romantic partner led to increased feelings of intimacy, so attuned are we to the movements of the bodies around us.

"People tend to synchronize with each other during ordinary activities, such as breathing, walking, and cycling," the researchers wrote in their paper, "Being on the Same Wavelength: Behavioral Synchrony Between Partners and Its Influence on the Experience of Intimacy." "Past research has indicated that such simple motor synchrony may inspire a sense of unity even between previously unacquainted interactional partners and have vast social consequences, such as heightened feelings of connectedness as well as increased cooperation and compassion." That is, simply moving your body in sync with another person, stranger or not, makes you feel closer to them. It's a powerful reminder of what our bodies need and what our bodies can do. It certainly kept me sitting in that hot-ass sauna.

It may surprise you to learn that I often think of my time as a ballet dancer with great nostalgia and longing. Fully unhinged, I know, but few things are cleanly demarcated into *pleasant* and *unpleasant*. It's true that my dance training involved psychological torture, the culmination of which nearly killed me. And true, I barely escaped the indoctrination to self-hatred that was impressed upon me by the sadistic teachers I worshipped in my youth, but! It is also true that something kept me in those rooms, with those people. Why was I willing to get into a hot mirrored studio in July with a group of other starving girls, to stand in front of an obviously drunk former New York City Ballet principal ballerina as she sloshed vodka out of her thermal coffee cup and mumbled slurred disappointments at us while we held our thighs apart from behind with our hands? She chain-smoked through the entire class and regularly made disgusted noises at my body, but nevertheless. Nevertheless.

Despite everything, there were things about ballet that were wonderful. Sometimes I catch a flickering glimpse of one of these old memories and my throat catches. Esoteric maintenance, like breaking in pointe shoes or hand-washing tights in the sink. The memories of the sheer ice-water adrenaline of dancing alone on a stage in a packed house or of the bone-deep satisfaction that came from a really good, productive, exhausting rehearsal—they all carry a certain sweetness to them, despite the bitterness of that time in my life. That said, when I think back on it all, most of my fondest memories of dance are communal. I miss dancing with other people.

Let me walk you through it.

Well before the start of class, people begin to trickle in from the street, lugging bags of gear. Inexplicably large bags. Dancers pack for class like doomsday preppers, but there is a logic to this. Ballet is rigorous, inflexible, demanding. Everyone who does it has a discipline fetish, and no one wants to be caught ill-prepared for their daily devotional. Out of the bags come leotards, backup leotards, wads of tights, multiple pairs of pointe shoes in various stages of mush (traditional pointe shoes break down with use), first aid supplies (tape, gauze, Band-Aids, liquid skin, hydrogen peroxide, Neosporin, Ace bandages, needles to pop blisters and lighters to sterilize them first), sweat towels, sewing kits, muscle salves, nail clippers, physical therapy torture toys, hair supplies, and warm-ups. Ah, warm-ups.

Warm-ups range from chic close-fit knits to absolute chaos pajamas, but everybody wears them. I loved that part, the start of class when we are all bundled in full-body sacks and thigh-high leg warmers and little wraparound sweaters. No one wants to face their body cold in the mirror, and warm-ups function like a security blanket, as well as a practical item, softening hips and hamstrings.

Before lining up at the barre, it's just puddles of bodies on the floor, limbs akimbo, running through Pilates hundreds (demonic abdominal exercises), rolling through hips and feet, pairing off in twos to roughly stretch each other out. It is very mechanical but has an element of spiritual practice too. Preparing the body for class is as important as preparing the mind. Ballet class requires personal focus, yes, but also a deep abiding respect for the focus and practice of those who worship with you.

Without prompting, each dancer selects a place at the barre (often the same one every day), the room quiets, and, often unceremoniously, the instructor takes their own space at the barre and begins to demonstrate the day's plié combination. Everyone watches intently. We are all about to suffer together, and it's important to suffer well.

Pliés are slow, rich knee bends, always first. Each day, they serve as an entryway to exploring what the body's feeling like today. It's an inventory exercise, it gets the blood pumping. Big, arching arm movements, bending forward and back, all these luxurious motions tucked between the stiffened, graceful raising and lowering of the body. We, the students, soak into the motion. The ritual takes over. Our breathing syncs up, and the air in the room changes; it is no longer a group of individuals with names and hopes and fears; it is a chorus of silent bodies, respiring and exerting together, more of a coordinated slime mold acting as one than collection of selves.

The exercises escalate in precision and vigor and difficulty. Gooey tendus, sharp dégagés, percussive frappés, lurid ronds de jambe, sensual, drippy *fondus*, sweeping, enormous grand battements. Each exercise asks something very specific of the practitioner, and the thread of sustained tension that links each dancer flexes and contracts in the shared exertion. With everyone doing the exact same thing, the mind clears; there is only *what step comes next* and *lift lift lift lift lift* and *scooooop* and all the other sounds that correspond to highly complex suites of mechanical motion. There is so much to remember always, and it hurts and it's hard and at any moment I can see a couple dozen other dancers doing the same thing, feeling the same thing, providing a pattern of possibility for me to shove my body into. This hurts, but everyone else is still going. This hurts, but no one else is grimacing. This hurts, but it also feels so, so good. They look happy, I feel happy, we are happy. Look at that beautiful arm; I am going to do that too. Look how she used the meat of the muscle of her feet to peel-spring it off the floor to her ear, look at where the force comes from; make your body a mirror and do that too. Behavioral synchrony at its finest, sweeping us away in a riptide of monkey see, monkey do.

It's not just an exercise technique. Behavioral synchrony is important to humans because it acts as a binding agent, a love spell of sorts. Moving bodies together makes us feel closer to each other, whether or not we are strangers. It also, rather curiously, and as I know all too well, has an effect on our pain tolerance.

A 2010 study from the University of Oxford found that behavioral synchrony is correlated with elevated pain thresholds. Plainly, it means that moving your body in sync with other bodies increases your ability to withstand pain. To study this, the researchers used members of a team sport known for its grueling, group-oriented physicality. They used rowers.

You see, like so many exertional activities, rowing a boat will give you a runner's high. So, before we get into the group dynamics around the all-mighty endorphin rush of exercise, let me take a brief moment to explain what, exactly, is going on in your body when you get that rush of good feelings after strenuous activity.

Specifically, the *runner's high* is a euphoric state that happens after a person runs or does some equivalent kind of endurance training. Aptly named, as this phenomenon has opioidergic mechanisms: the perceived euphoria from running comes from endorphins, which act on the body's opioid receptors. Yes, the very same ones that interact with heroin and other opiates. In fact, the word *endorphin* literally means "endogenous morphine," a handy portmanteau coined by Dr. Eric Simon in 1973. Studies have shown that the anti-overdose drug naloxone reverses the feel-good chemistry of a long run.

What the fuck?

This is a relatively recent finding, at least on the scale of human medical history. In the early 1970s, researchers from Sweden, Baltimore, and New York discovered that there are receptors in the human brain for opiates, such as morphine. But why would we have receptors specifically for the intoxicating euphoria of the poppy? Perhaps there is some homebrew morphine coming from inside the house?

As it turns out, yes: this is what is happening. If you were to skewer your head on thin wires, one passing through the bridge of your nose and toward the back of your head, and the other passing through your temples, the point of intersection would be at a pea-sized organ of enormous importance: the pineal gland. Master of the endocrine system, the pineal gland, when called upon, releases *endogenous opioid neuropeptides*: pain relief. These small molecules inhibit pain signaling and can make you feel, well, real good. Not unlike the externally produced stuff.

Using endurance athletes and a PET (positron emission tomography) scanner, researchers looked at the potential for opioidergic mechanisms to be responsible for the euphoria associated with running. That is, is the runner's high actually caused by endogenous opioids? (Surprisingly, there's not a lot of research on why running feels good to so many of its devoted practitioners!) And it looks like yes, the opioid theory of the runner's high is likely true.

That said, a separate study using non-elite athletes (and fMRI rather than PET scans) did not find evidence of running-induced euphoria in its participants, though the study noted other benefits of running for the participants. Could it be that the feel-good benefit to running must be built over time? I reached out to Dr. Jens Foell, of Chapter 2 fame, for his thoughts, and he said that potential explanation sounded plausible. "Could be that there's an immediate euphoric effect that can be too small to come up in these studies, and that the brain of a regular runner adapts to it in a way that intensifies the reaction, leading to the 2008 findings." It seems like it might be that the more you run, the more homemade brain drugs you get. Hopefully, further studies will shine more light on this fascinating subject.

As if the promise of homebrew brain opioids wasn't exciting enough, there is an even newer body of work revolving around exercise and the body's endocannabinoid system. That's *cannabinoid* as in cannabis; if endorphins are homemade opiates, then this is basically the weed version. Our cannabinoid receptors are what get acted on by various compounds in weed like cannabidiol (CBD) and the ever-popular delta-9-tetrahydrocannabinol, or THC. And it turns out our brain has a lot of these receptors. The current understanding is that we have more cannabinoid receptors than all of the other neuromodulatory receptors in the body combined! We create neurotransmitters like anandamide, which is involved in a host of bodily processes, including modulating our sense of pleasure and reward. Anyone who has unearthed a plate of leftovers from the fridge at 1:00 a.m. while baked as a cake knows what I'm talking about. Scientists have shown that blood levels of anandamide are significantly higher during vigorous exercise. So the runner's high might just be higher than we thought.

Knowing what we know about the runner's high—physical exertion can lead to the release of endogenous morphine (endorphins), and those endorphins can block pain signals—researchers constructed a study to test whether exercising as a group causes increased pain tolerance.

The Oxford research team found that when the rowers worked out together in a synchrony, they had a bigger endorphin surge than when they worked out alone. They note in the paper that the "heightened effect from synchronized activity may explain the sense of euphoria experienced during other social activities (such as laughter, music-making, and dancing) that are involved in social bonding in humans and possibly other vertebrates."

Just by virtue of the presence of my fellow dancers and the fact of our bodies moving as one, my pain tolerance went up in ballet class. There are other reasons, of course; nothing is ever as simple as "this" or "that," but, when I look back on my training, I see an unmistakable effect of groupthink on my own actions. Being in class with others and participating as the observer and the observed allowed me to push past my limits, both perceptual ("I did not think I could do this, but I can") and physical (feet so raw that they bled through my pointe shoes, torn tendons in my legs and ankles, a fractured lumbar vertebra, fainting).

I hurt myself a lot, and physical pain became a way to know myself. To know certainty. To find limits and test them and prove, over and over and over again, that I am tough enough to withstand this and resilient enough to recover. Pain is an intimate, wordless thing. Not soundless, not silent, but prelingual, before words come. Nearly impossible to articulate, and only really understandable to an outsider if you put it in terms of what might happen to their own body. As Elaine Scarry wrote in her book *The Body in Pain*:

Physical pain does not simply resist language but actively destroys it, bringing about an immediate reversion to a state anterior to language, to the sounds and cries a human being makes before language is learned.

So, what does it mean to be in pain with other people?

I know the people I danced with in a way that I will never know anyone else. There is deep intimacy in shared pain, in knowing someone else's pain in a very real and certain way. *I have a better understanding of what you are feeling because I am feeling it too.* I've felt the shared intimacy of physical pain in person. I've read about it quantified in scientific studies. I've seen it happen right before my eyes. And I think the bonding that happens when people experience pain together is due, at least in part, to the unknowable nature of pain. Or perhaps that is better termed as untellable. But if you move like I move, then you might feel what I feel, and that feeling is something that language struggles to define.

There is powerful bonding to be found when pain is shared, and humans know it. Think about all the rituals that involve pain, from religious flagellation to frat hazing. Think about how someone might use this phenomenon to facilitate intimate bonds between strangers. Think about the last time you did something in a really big group of people. Think about how I am about to run into the ocean with Rain.

"I am going to kick the ocean!"

Shortly after dawn, New Year's Day, I slink out of bed to brush last night's celebration out of my mouth and see that I have two messages from my friend Valerie, who has kicked off her New Year's Day in the Czech Republic with an icy plunge. I feel a shiver of benign peer pressure wash over me.

I watch the videos in a dark hotel bathroom, downing paper cups of tap water to atone for the whiskey I celebrated with hours prior. Rain is in the bed, no doubt as much a suffering fool as I, so I bring her a cup of water to rouse her.

"I got a message from Valerie," I say with the somber tone of a dramatic clergyman. "She did the plunge."

Rain sits up in bed with a squint. Or maybe a scowl. It's early, and the agenda of the day looms large: today, we are doing the 2020 Coney Island Polar Bear Plunge. Neither one of us is excited. In fact, it is this very activity that I dreaded the most when I pitched this book. Carolina Reaper peppers? Fine, whatever. Deeply intimate interviews with strangers about their bodies and their feelings? My tender heart loves that. But cold? Cold?? Have I mentioned how much I hate the cold? Have I sufficiently conveyed my primordial, bottom-of-the-bowels level of hatred of the cold in all of its forms?

I joked about the plunge killing me, and I hoped it was a joke. In the pantheon of my masochism—which has involved everything from having my body hair burned off while blindfolded in the woods, to getting mummified in plastic wrap and hung from the ceiling and beaten with a large wooden cheeseboard, to writing a book—cold ranked near the very bottom of my list of ways in which I was inclined to suffer for fun.

Our hotel is a cool mile away from Deno's Wonder Wheel Amusement Park, where the intrepid (stupid?) participants in this year's plunge would be queueing up for registration. Rain grabs a hotel lobby coffee, and I put cold orange juice in a paper cup, somehow still not grasping that I was about to be walking in the cold, never mind running into the ocean. Bundled and nervous (resigned? anxious?), we start our stroll to the beach. *I am going to kick the ocean*, I keep saying. *I am going to kick the ocean.* The cold wind immediately cuts through the elaborate, bespoke onesie I am wearing that's been made from a single, enormous sweater, and I draw my elaborate, goth Santa coat around me in an indignant huff. Dangling from my wrist is an unsettling plastic pig mask. Rain is stoic in her ankle-length leather duster, and her serene, determined look brings a sense of dignity to the comedy unfolding. She has one of those intensely structured faces, the kind of cheekbones and jawline that conspire to create the most intimidating of deadpan stares. Me, however, I have no dignity to speak of; I am still talking about how I am going to kick the sea.

The walk there is fantastic and cold and spooky in the way that beach towns get in the dead of winter; giant roller coasters slumber in the distance, surrounded by the empty space of absent revelers and unlit marquees. It's my first time, and I'm charmed by the utterly demonic smiling face that is plastered on everything from the rides to the signs to the T-shirts for sale in one of the lone shops open for business today. It's a white man with a toothy smile, ear-length black hair parted severely down the middle, pink-cheeked and red-lipped, wide blue eyes staring down from above. A face both ominous and chaotic.

The ghost town that is Coney Island in winter lays silent around me, save for a place to buy hot dogs and clams, and a steady, growing trickle of people walking to the beach. It's eerie, but it feels fun, illicit. Like a secret playground that only the strangest of revelers will get to frolic through. The morning of my polar plunge, I have only a passing awareness of the strange illness sweeping through Wuhan, Hubei Province, in China. A world away, it doesn't feel real yet. Months before, a new virus spilled over into humans, but the premise of it still feels small, remote, outside of This and Here and My Life. What I do not know is what is to come. How in hindsight, the empty streets of Coney Island were a harbinger of how the streets would soon empty for another, more infectious reason.

In experiencing this recent global pandemic, we see an unavoidable truth writ large: it is hard for humans to be away from other humans. Of course it is! Fundamentally, this is not what we do. In a time of crisis, we band together! Attack the foe! Rely on each other for comfort, aid, entertainment, physical touch, playfulness, physiological signaling—all gifts that we give each other. We look to each other for cues on how to act, our bodies ever ready to sync up, to move and make decisions together. Behavioral synchrony feels good! That is part of why it's so hard to be away from others.

But on the first day of the year, before COVID-19 staked its claim to the United States, before I could even conceive of the tide change to come, I am not separated from others. I am surrounded by others, mobbed by others, swimming through crowds of others. I stomp toward the beach, selfish and elated in my silly desire to do something *terrible but memorable.* I am excited to feel the power of the group. My cold limbs clatter with dread and adrenaline, questioning all the decisions in my life that brought me to this point. God, it feels so wild and reckless and fun and horrible, and it fills me to bursting like a poorly seamed scarecrow. I am myopic in my focus on the cold to come. I am focused to a degree that precludes worrying about anything else. It is a singular joy, the singular focus.

Rain and I stand in line, waiting to check in and get our wristbands and T-shirts. It's more procedural than I expected, though what, exactly, was I expecting? At least a couple thousand people did the plunge last year, so of course there would be security guards and plastic folding desks and clipboards and event staff. Some people are already in shorts, and it makes me want to die.

Finished with check-in and nothing to do except stand in the cold, we get in line for the bathroom. The public bathrooms are open-air, in that there is no door between bathroom and environment, and no climate control inside. The line is massive and snakes around. When I finally get a stall, I realize with resignation that wearing a giant sweater onesie means I have to get completely naked in my freezing bathroom stall. Totally starkers and nude on a cold toilet, I take a horrible hangover shit, the music of my bowels accompanied by the dulcet sounds of the person in the stall next to me, who is also just absolutely crapping their brains out. I feel a deep camaraderie.

Rain and I find lots of places to stand around, alternating between combing the beach for chicken bones and seashells and standing in urine-scented corners, away from the cutting wind.

There is a man on a microphone, instructing the crowd, narrating, and counting down the moments until it is time. NYPD cop boats float beyond the shore like brutalist angler fish, bobbing in the waves. The beach is packed, absolutely packed with revelers and watchers and swimmers alike; most of these people are warmly dressed, hands in pockets, heads covered with little knit caps. With the time to plunge approaching, Rain and I stand slightly off to the side of the main swarm, a section filled more with voyeurs and designated towel-holders than swimmers. The idea of being in the thick of it (a mere twenty feet away at most) makes me feel nervous, like I might get behind a slow Baywatch runner and be forced to extend my time in the sea. So I am on the fringes of the big group, which honestly describes my position in any large group activity. I want to enjoy a group, but never without a clear exit.

But I am here for the group, nonetheless. The baby who kept me in the sauna yesterday is now the elementary-aged child who, in their summer swimsuit best, just bolted into the ocean. I'd guess they are maybe eight years old, skinny as a beanpole, and they just absolutely scream their head off in the waves. A humbling reminder of what a baby I can be. Freezing in the winds on the beach, I weigh the pros and cons of fucking the procedural aspect of the day and just running into the ocean to get it over with, like the shrieking kid who just went in. Rain and I had arrived a little after eleven in the morning, stood in line for what felt like an hour but was probably only half that, and have been steadily losing heat for nearly two hours. Was I really going to wait until 1:00 p.m.?

Of course I was. That's the whole point. Personal Leigh wouldn't run in early because Personal Leigh wouldn't be here. Journalist Leigh wouldn't run in early because Journalist Leigh is here *specifically* to suffer en masse. My hands are tied. I have to wait. I tell Rain that she does not have to do this, that she would suffer only my increased respect for her if she backed out. This looks like a terrible idea! We are already so cold. But Rain shakes her head no. Better to suffer for a good story, a freezing memory, a test of self, a bonding moment.

When the time comes to strip out of my clothes, my fingertips and toes are already numb. I've been jumping up and down in the sand, doing sharp, vocalized breathing, but my angry chanting did little to warm my pitiful bones. The large man beside me glowers at my undignified display, but I don't have a single hot buttered fuck left to give. I'd pay twenty American dollars for a cup of warm water. I'd drink my own piss just to warm up my throat. Some stern slab of a man regarding me crossly is the least of my worries, what with the roiling gray Atlantic taunting me with her waves.

Close by on the beach, the day's emcee tells the swimmers not to run. I'm sorry, I did not realize I was going to have to *walk* into the freezing ocean? I raise an eyebrow. He also says to make sure we know where our stuff is, because "it's about to be a sea of humanity." The beach is packed, and still, the cold wind cuts through the crowd with ruthless ease.

"Fuck that, I'm running," I say, as I start taking off my coat. I haven't made eye contact with Rain for at least twenty minutes, not wishing to share the furious dread welling in my tear ducts. It's bad enough that we are here at all, and I don't think I'm ready to share with anyone the true nature of my discomfort, or reckon with just how miserable she is as well. It just seems more dignified to decline eye contact at this point. (As if I have any dignity left at this present moment!)

The second my coat comes off, the cuspidate wind announces itself instantly by slicing clean through my sweater onesie. I try to clear my mind, but I am furious. It's too cold, and my arrector pili muscles, already fully contracted, have little else to offer.

I unwrap my onesie and show my skin to the wind in what feels like an offering. I look like an angry plucked chicken. All around me, seeded through a crowd of parkas and scarves and balaclavas and gloves and knit caps and shiny thermoses are people just like me, nearly naked and about to greet the new year in the embrace of a cold, cold sea. I am going in with the first wave, then Rain, a practical consideration as to not lose our coats on a crowded beach.

The air hurts immediately, stealing the last cushion of warmth from under my armpits. I stand in the sand wearing a bikini, old black boots on their final outing, and a realistic pig mask hanging backward around my neck. (Armor made from absurdity is better than no armor at all.)

There is a marching band drumline, which I feel like is a nice touch, but I won't hear them until later, after I am out of the ocean. It's not that they weren't playing; it's just that as soon as I begin my jog down to the water, memory becomes a tunnel. What I remember is pruned by a harsh environment and my body's urgent attempts to withstand it.

For a second I stand at the water's edge, peering out at the vast gray expanse of the stupid decision I am about to make. I am ready. That's a lie. I will never be ready for this. I have read too much research by University of Portsmouth professor Mike Tipton, one of the foremost experts on thermoregulation in the human body. The thermoneutral water temperature for a naked human body is about 95 degrees Fahrenheit. Any temperature below that, and the body starts leaching heat. The high today is about 35 degrees, and I hear the water temperature is even lower. I think about the four Army Rangers who died of hypothermia during training exercises in a Florida swamp in 1995. Elite soldiers died from cold water. In Florida. What in the hell am I doing here? I know I won't be in the water long enough to get hypothermia, but there are plenty of other fun things to consider, like hyperventilation or a spiking pulse and blood pressure or heart arrhythmia. I could have a vasoconstriction-induced heart attack. As blood rushes away from my extremities inward to protect my vital organs, my limbs might just stop working, and I'll founder in the icy sea spray. About three thousand people die of "water immersion" every day worldwide, and many of those deaths are due to cold water temperatures. But I've been taking cold showers at home, as proof of concept. I think I'll be okay.

I try to remember a video that I watched about a Dutch guy named Wim Hof, who does this wild breathing before swimming in the Arctic or running topless up frozen mountainsides, as if I could mimic my way to his crafted cold tolerance, but nothing stirs in my frostbitten brain. And Rain is here, and so are thousands of other people who seem to be fine. Valerie has already done her plunge. I have a book to write and a slew of painful new experiences to experience. I can do this.

Running into the freezing ocean is an exercise in overwhelming one's better judgment. Nearly naked, fingers like little pink stones from two hours in a mid-Atlantic January morning, I hit the surf as water begins to pool in my boots. I am trying very hard not to feel it, but there's no denying the bitter chill enveloping my feet. But there's no stopping, no time for recoil. Only forward. Around me, I hear the screams of the other idiots. We jog and slosh forward, plowing into the surf.

The sensation of my bare feet in ice-water boots is strange enough to warp my attention, but by the time the water is up to my thighs, I cannot feel anything below my knees. It hurts like one million knives, straight out of the refrigerator, somehow everywhere against my skin at once. My mind is throttled with the memory of being made to sit in ice baths up to my chest when I was in physical therapy at a prestigious ballet school. That fucking rubber ducky that floated in the stand-up metal bathtub, filled with cubes of ice and water, and how all the physical therapists were mean little shits. But the memory flickers away as fast as it hits me. Turns out, there's not much room for thought in a brain this cold.

My legs are churning forward. It's hard to run in water this deep; sometimes it can lead to loss of muscle control. The cold hurts in the way that only cold can, and I feel the flesh of me becoming marble in the water. I am turning into the worst kind of statue, and I struggle against the pain in my screaming feet. They are numb. They hurt so much. I don't understand it. The icy water hits my genitals and I make a horrible gurgling, screeching sound like some kind of dying bird when the icy brine swaddles my labia.

I do the Kegel to end all Kegels and all I can hear is nothing, screaming, nothing, ringing, cold cold cold. You can hear cold, I swear it. Or maybe my bottom half was so freezing that my brain started making up new alarm bells to ring. I am filled with unholy rage. I am fucking furious, with myself, with the ocean, with the sky, with any deity within screaming distance. But once the water is past my pussy, I let the adrenaline take me away.

Still running, which at this point is a slow jog underwater, I make it out to chest height. Around me people are swimming, holding hands, howling, swearing, hollering, giggling, laughing, and splashing. There are breathless scowls and guttural howling and unhinged laughter.

I'm in the water up to my chest, edges of my vision useless, bones like bananas that have been dunked in liquid nitrogen. I am old refrigerator meat. I am a gremlin of the benthic seas. I am so angry that I cannot think straight, so cold that I cannot breathe right, so fucking stubborn that none of this is stopping me. I cannot feel my legs. I scream *OKAY!!!!!!* I count down *one, two, three!!!!!!* I take a deep breath and dunk my head underwater.

Brain freeze.

It's the worst brain freeze of my life, but there are no words for me now, just sick, devastating cold spilling down the back of my neck, organs hollerin' at the affront to their existence. There is only a singular want, all of my mind and body scrambling to do ONE THING, ONE THING RIGHT NOW and that's to get the fuck out of the Atlantic Ocean. Stumbling and clumsy with cold, I start toward the beach, legs buckling beneath me, arms flapping and useless. As if I could claw the air for warmth. The cold air on my wet body hurts, it hurts all over, and my skin screams the body electric in a key that only dogs can hear. I am laughing? I am laughing! I am howling with laughter! I am screaming like a banshee! I collect myself when I near the crowd and start my trek up the beach to Rain, shaking so hard that my eyes feel broken. Rain! Rain! Rain??

Cold and naked and baptized in the icy waters of a new year, I absolutely cannot find Rain. Suddenly too cold and serious to scream, I rush around, trying to find her, trying to figure out how to pilot my frigid body to warmth. These are panicked moments. I am so, so cold. I know I will be fine, but try telling that to my brain stem. Finally, I see her.

I scream her name.

She immediately rushes toward me, putting the camera down, but I am a maniac and I urge her to film this. Obliging, she aims the lens at me, and I realize I have nothing to say. I have never felt this much adrenaline in my life, I think. It's like taking a LOT of very shitty cocaine at once, and my heart feels like it's going to explode. I just start shaking my head back and forth, *no no no no no no no no no no no* and I tell her to stop filming. She's laughing, the crowd is buzzing, and I am shaking, smiling, relieved. People pour in and out of the water around us, each one in the midst of a private hell, together. Some sick fucks even seem to enjoy it.

I put my coat on over my wet body and wrap my hair in my wet towel. Now that I am back to watch our stuff, it's Rain's turn to meet the sea. She walks briskly toward the surf with a truly alarming amount of stoicism, slipping between the crowd of revelers like a cat at a house party. She marches directly into the ocean, and she marches back out, walking slowly and with purpose up the beach. I am howling with laughter, and the microphone catches all of it, my panting and giggling and ragged breath, just absolutely wrecked by joy and pain and stupidity and close calls and oh my GOD I am undone by all of it. Screaming, I ask her, "How was it, Rain!!!" and she doesn't reply. I ask again, annoying and totally blasted on whatever cold-water drugs my brain just made for me, and she says, simply, "Cold."

We laugh and laugh and laugh and complain loudly about how cold we are and join the exodus of wet revelers streaming away from the beach. We walk all the way back to the hotel on numb feet, hungry and freezing and high as kites. It was surely one of the stupidest things I've ever done, but now, and always, it will be a stupid thing that we did together. Humans love that.

Chapter Eight

THE ULTRAMARATHON

To put it mildly, my initial response to The Email was undignified. A friend had just returned from Oaxaca with mezcals impossible to get in North Carolina. I'd been sampling them and romping with her giant, geriatric deer of a dog under a nearly full moon until the soup and snacks buffering me from the abyss of drunkenness wore thin, and the alcohol pulled me into a pickled slumber, facedown on her couch. When I woke up later to be escorted home in the night, I had notification of an email from the creator of what is widely held to be one of the most challenging ultramarathons in the world, a last-runner-standing-style race: the Big Dog Backyard Ultra.

This email from Gary Cantrell, known to friends, family, and devoted running fans as Lazarus "Laz" Lake, wasn't unexpected. Six months prior, I'd secured press credentials to cover the 2019 race on-site at his property in rural Bell Buckle, Tennessee, a town that has a population of about five hundred people, per the 2010 US Census. When the email arrived, the race, affectionately dubbed "Big's" by participants, was in less than a week, and I was elated. It's a gruesome event, but on the surface it seems like a relatively placid affair: every hour, the runners have one hour to run a 4.1667-mile lap.

During the day, the lap is on a barely cleared trail that winds through the woods on Laz's property. At night, the laps are run on a paved back road. So far, so good. No rocky precipices, no arctic blasts, no sizzling in the frying pan of an unrelenting desert sun, no ravenous jungle insects—all features of other notable extreme ultras.

What makes this race so bad? Well, there's no preset distance to cover or time to finish. The race only ends when everyone has quit, save for one runner. Until then, they run laps in a sadistic war of attrition. The rules state that the runner must be back in the starting corral by the start of the next hour to begin the next lap. If you do not make it back in time, you are disqualified. If you cannot will your body from your chair to the starting pen, you are disqualified. The race goes until there is only one runner remaining. There are no extended breaks, and little pause for the necessities of life like eating, sleeping, and going to the bathroom. There are only the few stolen minutes between the runner returning from their lap and the ringing of the cowbells that signal the start of the next one.

When I drunkenly opened my email from Laz, I was expecting another message about the logistics of camping on-site, or where to park, or which porta potties were for runners only. What I got, however, made my cheeks burn hot with indignity.

> leigh,
>
> it came to my attention that your book is not related to sports,
> but rather is on sadomasochism.
> this is not a misconception that we want to perpetuate.
> like many sport there is discomfort involved,
> but it is a cost of competition,
> not an objective.
> i don't think that sort of coverage is appropriate for our event.
> this is a serious athletic competition.
> i should have been more proactive in checking on the nature of your
> book.
> i regret having not asked more questions up front.
>
> laz

Unmoored by ethanol and a brain filled with the cotton batting of sleep, I wept.

———

Getting to the race site requires parking 0.7 miles away in a field marked by a handmade sign and walking down a rural, tree-lined road. It's 4:45 in the morning, and I'm striding onto Big's property filled with hotel coffee, nervous excitement, and a feeling not unlike dread.

It's so beautiful here, where each lap will start and finish. A gravel road splits a field of tents, all softly glowing from within, tents that house the runners and their crew, who will serve them in their quest. The hushed bustle of race prep—sounds of rustling through ice-filled coolers, the susurration of bodies stirring inside sleeping bags, the incessant crinkling of plastic wrappers—belies the radical suffering to come. Low voices and respectful headlamp use float through the air. Laz, looking every bit like a chaotic Santa Claus, hovers near the starting corral, his red flannel shirt and red beanie and long salt-and-pepper beard drawing eyes to him with ease. He's the mastermind behind the race, a known and venerated sadist in the world of endurance racing. (He's also the brains behind the Barkley Marathons, a nigh-unfinishable race through the woods with no course and lots, and lots, and lots of thorns.) He cracks jokes in a sharp, nasal voice, his eyes peering through thick glasses, seemingly bemused at all the runners before him who have trained so hard to meet their limits here today. In spite of everything that is coming, or perhaps because of it, there is a certain sweetness to the predawn courtesy around the campsite. Presumably, no one has been up long, and the sun has yet to begin her pastel creep into the sky. There is just so much activity to come. My stomach is a mess.

It's in the upper forties and breezy, and the beam of my headlamp picks up the swirling particles of a faint, low fog. I prefer to hike without a headlamp at night, a habit I picked up doing bat research in Costa Rica, but it's unwise to take that risk as I shuttle between my car and my tent this morning. Here in rural Tennessee, the light of a moon on the thick side of waning is plenty bright enough to navigate the countryside, but there are cars full of Australian runners looking for directions and devoted spouses organizing support crews, so I flick on my lamp when I hear the rumble of an engine.

But there is a payoff here too: at the edges of my beam, where the light scatter is overtaken by darkness, the ground appears covered in tiny, glittering opals. Spiders.

Their eyes catch my light and return it to me, hundreds of them embedded in the forest floor, which has been freshly dusted with autumn's first carpet of leaves. In the midst of the massive Anthropocene extinction event, there is delight to be had in such thriving populations. Many of the trees are still heavy with leaves, but these forest giants, black walnuts and white oaks and maples the height of county courthouses, are midway through a spectacular show, burning red with anthocyanins as the first blush of highlighter orange seeps through their leaves, flush with carotenoids.

My tent is on the edge of the congregation, closest to the forest, and I can smell faintly the musk of a fox mingling with someone's hot coffee, though it's hard to smell anything over the very specific aroma of a used camping tent, never mind a small sea of them. All the runners are picking up their race tags and ankle monitors. The sky has turned from navy to lavender, and there is enough light now to silhouette the trees. I warmed up earlier with a brisk two miles of loading in gear, but stillness has stolen what heat I generated through activity.

It seems likely that the schadenfreude of watching this race will help with my menstrual cramps, which delighted me by waking me up twenty minutes before my 4:00 a.m. alarm clock this morning. My uterus feels like it is trying to contract into a sliver and slip out of my cervix and I'm bleeding heavily. But nothing I am feeling now could ever hold a candle to the bewildering display of human suffering that I am about to witness.

It's 6:38 a.m., and Laz is spray-painting a large rectangle onto the gravel around the cadre of runners gathering for the start of the race. The smell of aerosolized paint and propellant is overpowering. A tall, lanky man with shaggy blond hair and tiny white running shorts saunters up to the starting corral with a half-eaten sausage biscuit in his hand, thoughtfully chewing. The crowd is mostly white men in their thirties and forties, but there are some older folks, some women, a few people of color. Before me mingles a cluster of athletic gear, sinewy legs, well-worn shoes, light jackets, the pungent scent of human excitement. With his reedy voice, Laz barks his final words of wisdom before the trial begins:

"Don't poop in the woods!" A gentle chuckle rumbles through the crowd. "The dogs will roll in it!"

At precisely 6:40 a.m., after a snarky cheer from the race's two *jeerleaders*, and right as the crest of dawn mounts the soft hills around us, the starting cowbell clangs a few times. With little fanfare, the runners are off, jogging out onto the path under a giant, inflatable arch that says FINISH. In a few hours, a white sheet will be draped over the archway with an addition, changing it to read THERE IS NO FINISH. Grim humor abounds in Laz's famed races.

The seventy-two runners who have qualified and desire to be here today shuffle off down the road for the first part of the course. The trail in the woods is a cool 4 miles; they must add on a 0.167 miles of road running each lap to meet the course requirement of 4.167 miles a loop. (Chosen so that after 24 hours, the runners will have run 100 miles exactly.) It's exciting. Almost as soon as they leave the starting corral, the runners trot down the road and come back through the starting area and head into the woods, the early dawn light filtering down on them through what remains of the leaves. It also means that all runners have to run by their chairs before entering the woods, a trap that will end many a race later this weekend. We cheer them on at each opportunity, and with the first lap underway, we begin the practice of keeping watch. We wait for them to come back.

I grew up in a family full of runners. My father, his brothers, and my grandfather were all runners, and my dad continues to this day. My knees were already shredded from ballet by the time I was in middle school, which, notably, did not stop me from dancing but definitely kept me from running seriously. My overdeveloped quads pulled too hard on my growing knees, and my patella tendons remain pointy and ossified from the strain. This excused my inability or unwillingness to carry on the family tradition. That is to say, while my particular preference for painful hobbies tended more toward more artistic pursuits, I still grew up steeped in the lore of the runner.

My grandfather was an early adopter of leisure jogging in Florida during the 1950s, and the sight of his regularly and seemingly aimless running was so concerning to his fellow townsfolk that the local newspaper ran an article about him to reduce the number of emergency calls coming in regarding that strange running man. My father hiked the entire 2,020 miles of the Appalachian Trail in one shot, from Georgia to Maine, at age sixty-three, as a retirement gift to himself. When my mother had both of her knees simultaneously replaced at age sixty-one, she healed so quickly from the surgery that she was discharged directly from the hospital to home, no longer qualified for the inpatient physical therapy as planned. My sister is a long-distance swimmer, powering herself across whole lakes with nothing but her steady breath and broad shoulders to drive her. My child ran her first 5K at age seven while I was covering the race at Big's and called me to relay that she was elated to have sprinted over the finish line like I taught her just the year before.

What I am saying is that I am part of a lineage of hearty, active people who are neither relaxed nor relaxing. The scattered *Runner's World* magazines, Prefontaine biographies, and wrinkled, framed race bibs around my home easily held captive my childhood imagination. At some point early on (and I know it was early because I can't remember not knowing it), I absorbed the legend of the marathon, a mighty race that I knew my father had run. I understood precious little else about *the marathon*, other than this story. The version I grew up with was that a Greek runner was sent 26.2 miles from Marathon to Athens with news of a victory, then dropped dead on the spot. It was surely a disconcerting story to hear about one's father's hobby, but I remember feeling awe more than fear.

I kept that story, unquestioned in my heart, for decades, the way childhood legends and misconceptions defy fact-checking deep into adulthood. (Like how my partner was forty years old when they learned that ponies do not grow up to be horses.)

It wasn't until I was at Big's that it occurred to me to look up the history of the marathon. Turns out, the story that I and so many other people know is, well, wrong.

Accounts vary, but according to the Greek historian Herodotus, during the battle of Marathon in late summer 490 BCE, a runner named Pheidippides booked it from Athens to Sparta to ask for help, not to trumpet victory. He then turned around and ran all the way back to Athens to deliver the bad news that help was not on the way. Here's where the story gets more interesting: not only did he NOT DIE, as the legend holds, but the distance he ran was well over the 26.2 miles of the modern marathon.*

No, in his historic trek, Pheidippides ran 280 miles because it was his job. (This distance is also, as it happens, almost the exact length of the course record set here at Big's: Johan Steene's astonishing 283-mile trek, assisted by runner-up Courtney Dauwalter in 2018.) Pheidippides was a day-runner, also known as a *hemerodrome*, an Athenian military courier. His job was to run communications between cities, and quickly. The extreme distance and mountainous terrain between the two Greek cities rendered equestrian travel impractical. With today's amenities like telecommunications, energy gels, and cushy shoes, it's easy to think that running ultralong distances would be a development of the modern age, but that's just not true. In fact, humans are so well suited to running that the emergence of the ability millions of years ago might have been one of the things that made us who we are today.

Let me explain.

It turns out humans are uniquely suited to endurance performance in a way few other mammals are. Some researchers hypothesize that the act of running might actually have helped make us anatomically modern humans. About 2.5 to 3 million years after humanity's ancient relatives became bipedal, something happened. From what we can surmise from the fossil record, a suite of physical characteristics began to emerge that differentiated the earlier *Australopithecus* from the later *Homo*, changes that paved the way for early humans to look, well, much like we do today.

The shoulders dropped and decoupled from the neck, allowing the body to move separately from the head. The forearms got shorter while the legs got longer, and the connection between the pelvis and the spine grew more robust. The bones of the feet shifted and squared up, while the surface area of the ankle, knee, and hip joints got wider. A huge ligament showed up, from the back of the skull down the spine. We grew butts.

---

* Some accounts still hold that he did die after the second run to Marathon and back to Athens. So he ran 150 miles over two days to Sparta, 150 miles over two days back to Athens, then over 50 miles from Athens to Marathon and back again. He could be forgiven for keeling over dead after a run like that.

And here's the thing: bipedalism alone cannot explain the physical changes from *Australopithecus* to *Homo*, never mind that these changes took place millions of years after we stood up. So, if getting up on two legs didn't cause these changes, what did?

Compared to other mammals, humans are not particularly fast sprinters. This made it tempting for scientists to initially discount running as a significant factor in human evolution. However, the simultaneous emergence of a whole group of anatomical features might best be explained by reconsidering that premise. Running fast is not the only way to run. What about running long?

In their paper "Endurance Running and the Evolution of *Homo*," researchers Dennis Bramble and Daniel Lieberman argue that we evolved from other apelike primates because natural selection favored those who could run and, specifically, those who could keep running. Their work puts forth the very compelling theory that the ability to run long distances shaped the way humans look today.

Watching the runners at Big's trundle in from their first 4.167 miles, it's easy to see the changes that differentiated *Homo* in action. With their long legs, short forearms, and springy tendons, humans have more efficient and stable body mechanics for running, especially compared to the long-armed, short-legged *Australopithecus*, with their permanently shrugged shoulders. Our shoulders move freely from the neck, providing balance, and things are further stabilized by a flatter, rounder head and new neck tendons. Each runner uses their forward-pointing big toe that, years ago, came around to make a square foot, which is great for push-off; our wider joints and a bigger heel bone make running easier on the body, functioning as improved shock absorbers. And then there is the wonderful human butt. Critical for stabilization, butts—important for running, but not really walking—keep us from falling over on our faces when we pitch forward to run. No butt, no running.

And it's not just body mechanics that make humans good distance runners! The new, long and lean *Homo* body type that surfaced millennia ago meant that, relative to the volume of our insides, there is more surface area for heat evaporation off the skin. Skull changes and the position of our blood vessels further help to cool large quantities of blood, and fast. A lack of body hair helps too, allowing heat to escape. And, blessedly, we are covered in little glands that squirt salty water onto our skin for the purpose of evaporative cooling.

You see, unlike most other mammals, humans can sweat. We do not need to stop and pant to expel heat, which is rare. We can breathe hard while we exert ourselves, and our breathing is not tied to our gait. In addition to all the other ways humans are adept at endurance running, cooling by sweating, rather than panting, means that *humans can outrun almost any animal on the planet.* Not by sprinting, of course. But through our relentless capacity for endurance. On a hot day, a human can even outrun a horse, and in a distance as short as one marathon. I wonder if the race I am about to see is long enough to kill a horse. (It is.)

But why were early humans doing this? For food, most likely! Being able to run long probably made us better scavengers, capitalizing on the presence of a column of vultures over an animal carcass in the distance. And then there is the rather harrowing practice of persistence hunting, when humans coordinate to run down prey animals over long distances, their pursuit exhausting the animal to the point of collapse. Unable to rest enough to pant and cool themselves, and eventually too weak to fend off the hungry primates that ceaselessly track them for hours and hours, these much-faster prey animals are ended by slow, relentless joggers. I can't help but think about the slasher movie trope of the killer who is always walking, never running, and yet still manages to catch and kill each victim.

As I watch this phenomenal race unfold over the course of an entire weekend, it's easy to imagine ancient humans, ceaselessly running down their gasping prey.

Most of the runners are back from their first lap now. The trail they run on during the day is mostly cleared, dry, full of hills to charge over and roots to trip on, and it ends in a stretch of gravel road, bringing them back to the start, flanked by tents filled with helpers. There's a thin, wiry man who appears to be in his late sixties and yet also ageless, seemingly impervious to the cold as he sets up his crew site in belted khakis and a neatly tucked-in T-shirt. For context, I am wearing thermal socks, thick leggings, a thin, long-sleeve thermal shirt, black hoodie, knit cap, and a battery-heated jacket. My nose is still cold, and I am humbled by the many, many hearty folk who bound around the race site in light jackets, invigorated by the morning air. Most of the crew people are runners themselves, and solely on the basis of wardrobe they appear to be dramatically tougher than I am. Minutes earlier, the fastest runners of the bunch had streamed into the tents with twenty minutes to spare, but now the more experienced runners are sneaking in with just minutes left before the starting cowbells of the next lap. Pacing is everything at Big's. Counterintuitively, no one who has ever come in ahead of the pack on one of the first sixteen laps has ever won the Big Backyard Ultra.

And for that matter, neither has a woman.

But all that might be changing. Katie Wright is here, the first woman ever to win a last-runner-standing race in her home country of New Zealand. And last year, American powerhouse Courtney Dauwalter almost won Big's, making it sixty-seven hours and 279 miles before dropping out, a remarkable assist to Johan Steene's record-breaking win. Laz wants to see a woman win it, and there are many contenders here today. It's time, he tells people. It's time.

Two crew members near me express quiet bewilderment at the runners who come in fast. "The people in the back know what's up," I hear as I slink around the starting line, a quiet voyeur in the midst of very focused athletes.

Closest to the start is a huge, open tent, the kind used by catering services for outdoor weddings, filled with fabric and wire folding chairs. The chairs sit empty, waiting for the runners and filling up for brief moments before emptying anew. It is one of the hardest things about this race, the siren call to *stay here in this chair* that builds as willpower wanes. But it's early yet.

Even though it's only the end of the first lap, many of the runners make a beeline for their particular chair in the sea of all the chairs, put their feet up, and close their eyes. Others run directly toward a mat on the ground and drop fast to lie prone under a blanket, to rest as much as possible in the four minutes remaining between *now* and *start*. But not all of today's contenders take such a conservative approach. Other runners are up and walking around, springy, stretchy, asking their crew members for particular items or trying to figure out how to set up their chairs closer to the starting line. Closeness to the start is important in a race that can go on for days; every step adds up, takes energy, and, eventually, hurts.

Guillaume Calmettes, the winner from 2017, is camped across from the Big Tent with another runner; the two sit behind their respective French and American flags. Near them is Canadian runner Dave Proctor, with his flare-orange shirt and cowboy hat and serious-yet-cheerful demeanor; he runs for Outrun Rare to raise money for rare diseases, the parent of a child afflicted with one. He also holds the world record for fastest 100 miles on a treadmill, at 12 hours, 32 minutes, and 26 seconds. That's nearly seven and a half minutes per mile for 100 miles! Near them is American Maggie Guterl and her crew, the latter of which happen to be the race's *jeerleaders*. Think cheerleaders, if their job was to gently demoralize you into quitting through exuberantly delivered and very depressing chants, like this gem from 2018: "Q-U-I-T! If you just quit, then you'll be free!"

There is a flurry of activity as people rouse and fuss with their jackets and socks, chit-chatting about the course in low, hushed voices while having bites of snacks and sips of water. Then, all at once, with very little time to spare, the congregation of runners returns to the starting corral, the clock counts down, the cowbells clang, and the runners are off for lap 2.

Then we wait.

───────

While the evolutionary legacy of running long distances serves as an elegant reminder of our shared human history and innate physical capacities, the shorter version of how we got here today is that one time a dude showed up to a horse race without a horse. Seriously.

In 1954, a small group of friends decided to ride a hundred miles from Tahoe to Auburn to prove how great their horses were. This was planned as a one-off event, the stuff of fireside scuttlebutt, but once other people saw that it was possible, they wanted in. Such was born the Western States 100, known today as the Tevis Cup.

In the early 1970s, a rider named Gordon "Gordy" Ainsleigh entered the race on his horse Rebel. Gordy was a big man, over two hundred pounds, which makes tough work for even the most robust endurance horse. As such, instead of riding his horse down the steep desert hills at a walking pace, as the other riders do, Gordy jumped off his horse and the two ran down together. Gordy had been a runner since childhood, and in a race as hilly as the Western States 100, he used that skill quite a bit. So, in 1974, Drucilla Barner, secretary of the Western States Trail Foundation, suggested that Gordy run the race himself. On foot.

Gordy declined. In a wonderful profile of the man who would become the father of modern ultramarathon culture, Karen Given of Boston NPR station WBUR reported that the reason Gordy finally decided to run the race was because he procrastinated on getting a new horse to ride. Without a mount, Gordy decided to give running a shot.

"I got to the forty-mile point, and I just said, 'I'm never gonna make this,'" Gordy told Given. The air temperature was 107 degrees that day. "It was so hot, I couldn't even focus. So I got to the point where I said, 'Well, what can I do?' and this voice from inside me said, 'I can still take one more step.'"

Shortly thereafter, his will—and his body—nearly failed him. If not for the grace of a dear friend waiting with salt tablets and water at mile 43, Gordy's attempt at greatness would have stalled out in the desert. He'd only stashed ten bottles of sports drink along his route, which wasn't enough for a hundred-mile trek through the heat. Rejuvenated by electrolytes and the milk of human kindness, Gordy went on to finish the race, performing a diving somersault across the finish line.

The next year, another man attempted the run, but quit at mile 98; the following year, in 1976, Ken Shirk, known as Cowman, completed the race in 24.5 hours. In 1977, fourteen men started, with three finishers. In 1978, sixty-three runners started, and thirty finished. The snowball effect was underway.

An ultramarathon is defined as any race longer than a standard marathon, 26.2 miles. Participation in ultramarathons has increased over 1,000 percent in the last decade. Steve Diederich, who helms the Run Ultra website and exhaustively catalogs participation in the sport, told *The Guardian* that in 2006 there were 160 ultramarathon races around the world. In 2018, there were over 1,800 races listed on the site. In an April 2018 article about the sport, Adharanand Finn also cites statistics from *UltraRunning* magazine: in 2003, around 18,000 people finished an ultramarathon in North America. By 2016, that number had ballooned to 105,000.

Back at day one of Laz's endless race, the runners are returning to the start, finishing up their second hour of running, a little over eight miles logged and ready to snack. The scene is nearly identical to the one that preceded it forty-five minutes prior. The runners shuffle in, head to their respective mats or chairs or chatting spots, fuss with their gear (VERY IMPORTANT), and eat. I note the importance of paying close, detailed attention to clothing and rubbing because a minor irritation on day one can turn into a full-blown horror show after days of racing. Saturday morning's light nipple chafing will easily become Sunday morning's bloody nipples will become Sunday night's blood-soaked T-shirt; light, damp toe rubbing for the first forty miles could be an entire foot's worth of Cronenbergian Bubble Wrap blisters and weeping hamburger meat by mile 140. Little problems have exponential potential to fuck you up when all you are doing is running, shitting, hydrating, pissing, eating, and running again.

The runners head off on lap 3, and again we wait. They come back again and again and again, and we wait in the warm Tennessee sun, ready with cowbells and cheers and jeers and plastic folding tables full of snacks. Nearby, a speaker blasts a steady stream of adult contemporary music, like Journey ("Love Will Find You") and Eric Clapton ("Cocaine"). I have a very singular experience involving a distressing, menstrual-cramp-amplified bowel movement in a sauna-like porta potty to the dulcet tones of "Livin' la Vida Loca" by Ricky Martin. I feel a twinge of camaraderie for the gastric horror show that will no doubt plague so many of the brave runners here today. Running long can be hard on the guts.

Hours pass.

The day marches on and late afternoon sun cooks the ground. The runners have been running for nearly nine hours now. "Everybody has different questions," says a runner nearby. "Maybe a different day I'll have different questions, but not today." I hear him tell a friend that he got what he wanted. At first, I wasn't exactly sure what I was hearing, but then it hits me: He's quitting. The man leaves the race so softly and quickly. One minute he was in it, refueling, then, just like that, he was done. Three runners have dropped so far today. Laz, who is prone to prosaic poetry, posts hourly updates on his Facebook page for all the eager fans at home.

there is a certain calm over the field this year
everyone knows this is going to be a long haul . . .
and everyone plans to be there when the game really starts.
so for now we settle for the slow accumulation of punishment to start taking effect.

Ah yes. The slow accumulation of punishment.

The sight of Canadian Dave Proctor's cowboy hat springs around the corner, calling out a hello to the crew people standing watch. Pleasantries are exchanged. Dave is implacable, smiling, *jaunty*. He has just run over thirty-seven miles, his ramrod posture and turbocharged legs propelling his slim form through space with seeming ease. Sometimes, when he is running long, he likes to make phone calls. He fills me with awe and maybe slight terror. I think about persistence hunters running down big prey on foot, and I picture Dave, trucking along with the confidence of knowing he doesn't have to stop, his smiling face the last thing an exhausted antelope sees as it gives up forever.

Not everyone looks so good. With one marathon down and the second one well underway, many runners are starting to look like I would assume they feel, which is bad. I overhear a crew member tell his runner, a tall, slender man named Shawn Webber, with mile-long cantilevered legs, that "running today is about running tomorrow," and a full-body shudder rips through me. It is one thing to know, abstractly, that this is a very, very long race. It is another to mill about a backyard in rural Tennessee for an entire day, only to realize that the race hasn't even really started yet.

By the start of the eleventh lap, two more of the runners have quit. Nightfall is approaching, and the pack has been running since dawn.

Laz says that the hardest part of the course is the walk from the chair to the starting corral. Every time a runner returns, they get to sit down. They have to decide to get back up again and start running again. Ten hours of running and forty miles logged is nigh unimaginable to me, and I keep having to forcibly return my thoughts to reality as I contemplate the eleventh lap in progress. The standards are different here at Big's, and I don't want to lose my sense of awe at what I'm seeing. And yet, forty miles is peanuts compared to the 283 miles that Johan Steene ran last year. Watching today's race unfold in real time, it still seems incomprehensible. Before the tenth lap, Laz resprays the ground and the smell of paint clings to the wet Appalachian air, mingling with the redolent bouquet of Icy Hot, body odor, and commercial deodorant.

I feel like now is a good time to mention that there is no prize money to be won here.

It used to be that people could sign up for these impressive physical slogs on race day, but no more. Many of today's races require lotteries and qualifiers and still sell out in minutes. To get to Big's, runners have three ways to qualify: they can win a Golden Ticket race, they can do well at an affiliate race, or they could get a personal invitation. For the 2020 Big's Backyard Ultra World Championship, there are thirty-six Golden Ticket races in twenty-five countries around the world. Winning one of those guarantees the runner a place at Big's; other top contenders from other affiliate races can earn their slot in the starting corral, but it's not a sure thing.

And these races aren't cheap either. Although some have low- to no-cost entry fees, even the cheapest race still requires travel and accommodations, multiple changes of clothes, quality shoes and quality socks, underwear, headlamps, a well-stocked cache of medical supplies, camping equipment, food for runners and crew, lots of water and electrolyte beverages, and a job that pays sufficiently to afford enough free time to train long and hard. And say you get good enough to start eyeing one of the more iconic, standard-form ultramarathons? You know, the kind with a finish line?

Well, the Marathon des Sables, a 156-mile race through the Sahara Desert that has had two people die while running it (so far), sells out in minutes every year and has a 2020 entry fee of 3,170 euros, or about $3,470, and that doesn't include the formidable airfare required to get you from wherever you live to, well, the middle of the Sahara. Runners who complete the Marathon des Sables are allowed to download a certificate of completion three months after the event. That's the prize.

Why, then, do people run these?

"A lot of things you do in life are easy, so maybe there's some kind of human nature [in] yearning to suffer," says Maggie Guterl. "Maybe it's like childbirth. You forget about the one painful part because the rest is a reward." She mentions the endorphins that come, the sense of satisfaction from pushing through pain. "Would it feel as important if you didn't suffer at all? I don't know. Maybe not."

Darkness falls, and the runners switch to the road. They've been running twelve hours straight on the hilly trails, but now there is something new to do. It's a welcome switch for some; for others, the newly jarring terrain and enveloping darkness make for tough work. They head out onto the pavement in their headlamps to grind it out until sunrise, if they can make it that long. It is raining now, and they've been running for fifty miles.

During the night laps, some of the sixty-four remaining runners stop by to warm themselves by the fire that is now roaring away in a cinder-block pit. Around the fire, competitors who have dropped out mingle with crew people, swapping stories of previous races and bodily horror, drinking beers and eating pizza, awash in the glow of being finished. Besides gawking each hour as the runners come through, there is little to do here. I pace a rocky section of earth, having already logged a laughably mild seven miles of wandering back and forth to my car, where I stashed books and snacks that would require light physical activity to retrieve. The runners keep going and going and going and going, and I pace and wander and sit with my feet stuck out of my unzipped tent, chewing turkey jerky and taking pages and pages of notes I will never use.

By the time lap 17 begins, it is nearing midnight. The runners have been going since dawn. They haven't even run a quarter of the distance covered last year. Laz continues to post his hourly updates, chronicling the event on his public Facebook page with short poems like this one:

even with 58 runners still on the road
there is silence.
just the sounds of shuffling feet.
adding up the miles.
one at a time.
the finish is still days away.
the finish . . .
there is no finish.
everyone is alone with their demons.
trying to survive the night
fearing the endless road
fearing the return to the trail.

there is nothing to look forward to.

no target to set your sights on.

just survive this hour.

so you can do it again next hour.

this is when the will must be unyielding.

the heart unwavering.

this is hell.

this event is hard.

Midnight passes, and the chatter around the fire turns to 2017 winner Guillaume Calmettes, who came to Big's that year to earn a place in Laz's other hell race, the Barkley Marathons. (Win at Big's, and you get to attempt the nearly unfinishable race in the brush-filled Tennessee mountains, near the prison from which Martin Luther King Jr.'s assassin escaped and was subsequently captured.) The Frenchman is a data scientist who speaks through a wide smile with charming enthusiasm. He won in 2017 with a race of 245 miles. But the runners around me aren't talking about his race from two years ago. They are talking about his leg muscles, which appear to spill over his kneecaps from above, as if his thighs, bored with their intended purpose, have also decided to take on extra work.

As if on cue, Guillaume walks out of the darkness and up to the fire.

"How do you do that?!" a fellow runner asks him earnestly. "What's your workout?"

Guillaume laughs. "You do not choose the body," he says, drawing out the words in his French accent, leaning away from the heat of the fire as smoke billows skyward. "The body chooses you."

The conversation turns to fawning. *You look so good out there! What a race you're having! Great form, great form!* Guillaume calmly puts his heel up on the cinder block that contains the flames and smiles wryly at the faces around him. "Sure I do," he says. "Or I am faking it? That's the name of the game, no?" Everybody laughs. He really does look good out there.

Two laps later, Guillaume is out of the race, shocking everyone. However, unlike some of the other runners, who find their limit, pack up, load out, and leave soon after their race is lost, Guillaume stays at camp. The sun is about to come up on day two, and the race is about to officially begin.

Perhaps a perfect avatar of the physical ordeal occurring in each of the sweaty bodies before me are the temporary toilets that dot the property. Which is to say, the porta potties here are already hurting. Whereas usually your shit disappears into a dark blue lagoon of chemicals that surely must cause cancer in the state of California, the ones here in Bell Buckle have no such abyss. Feels real fucking communal. It's so cold that my used tampon steams in the predawn darkness. Someone has placed a miniature scented candle on the floor of the porta john, and its tiny flame flickers in vain against the eye-watering miasma of the shit box. It is simultaneously terrible and miles better than anything those runners are feeling. I feel a rush of sympathy.

By morning, there are forty-three runners remaining, forty-three souls willing to go back into the woods and greet another hour on the trail. It will be hard, though. It rained during the night, covering the trail with slick leaves and even slicker mud. People fall at this race all the time. Most of them just get back up and keep running.

Maggie Guterl had a rough night, with knee swelling that started early, around mile 20. She was resigned to it only getting worse, initially keeping it secret before fessing up to the mounting peril in her leg. But with the help of her crew, and her stalwart resolution to get through some intense bouts of sleepiness through the night, she came back to the trail on day two, eager to run. "It's the kind of change everyone seems like they look forward to," she says, of the track switch that happens every twelve hours. "You're off the boring road, or you finally have a break to open your legs and run." The change between road and trail, she says, is welcome. "By the end of each section you're like, 'I can't wait to go back.'" With the sun in the sky and weather that can best be described as "ground cloud," she is excited to return to the trail. Dave Proctor is still churning along at a fast pace, coming in early alongside Swedish runner Anna Carlsson, who runs like the Terminator and grabs onto tree branches to sling herself around tight corners in the forest.

Katie Wright is still in the race too. Her mom is there, and we often pace in nearby circles. The previous day her mother told me that she worried about her daughter's body, "but I'd never tell her that." Wright drops to the ground and lies in repose like a beautiful seraph between loops, coming in and getting immediately horizontal on her mat with her eyes closed in angelic stillness. She has a tenacious, plodding gate. Both her legs and her demeanor suggest to me that she could outrun a horse, and then probably another one too. In 2017, she ran around the entire 870-mile coastline of Wales to raise money for charity. At night during her coastal trek, she told me that she had to sleep with her legs up on the wall because they kept kicking. After so much running, they didn't know how to stop moving.

Each time the runners come back out of the woods, they are muddier and sweatier, though some look worse for wear than others. There's an Ecuadorian runner named Andres Villagran who has some kind of nightmare unfurling around his right knee, which is now twice the size of his left. Yesterday, I saw him almost miss a lap start because he was messing with a patella strap; today, the back of his knee is blooming like an infected bite site in a cheesy zombie movie. Purples and blacks and dark reds streak out from around his knee strap, and his face is pale and clammy. It looks like there is some evil thing inside his knee, spreading. It's nothing short of gruesome, and, even in this crowd of stoics, there are murmurs of concern and horror at the sight of it.

In a rare move, Laz advises that Villagran stop the race. But Villagran wants to continue. He keeps making it back in time, and, importantly, his sole crewman is his father. Were that not the case, perhaps there would have been more intervention to assist Villagran in ending his bid at being the last runner standing. As it stands, Laz grudgingly accepts that father knows best and logs a concern for Villagran's health. The entire point of the race is for runners to push themselves beyond anything they've endured before. But even so, Laz and others are worried about permanent injuries. Where are the lines between informed, adult consent and a form of torture that no one should be encouraged to dabble in? There is no one answer. Everyone here has to grapple with the enormity of what is being asked of these runners. As a spectator, it's hard to keep from dissociating in the face of something so . . . surreal? I know that these people started running yesterday at dawn, and that they are still running today, and that I've been watching the whole time, and yet. And yet.

I am not sure I believe it. A warm Sunday unfolds from under the cold, wet morning: picnic weather for a dilettante, but an oven for the runners, who trudge on in their ceaseless parade of pain, only stopping to figure out little things like socks and snacks and what kind of tops to wear. It's the attention paid to these minutiae that makes their extended suffering possible, and the crews tend to them like patient worshippers of some lost old gods.

The crews are indispensable. There are so many problems to solve when pushing a human body so hard for so long. What and how to eat; which muscles to stretch and which muscles to rub; what to wear. After about fifty miles, the body stops being able to thermoregulate very well, and it's up to the crew and the runners to endlessly problem-solve the question of what to wear.

It is hard to watch this small cadre of weary bodies and not wonder at length about the battering they face. After running so far, it's not uncommon to piss blood merely because of the repetitive action of the foot striking the ground, the accumulating, unending pitter-pat force of heel-to-earth enough to shred some hustling red blood cells. It's also possible to run so hard and long that the relentless jiggling of the bowels, combined with an exercise-induced lack of blood flow to the area (*thanks, muscles!*) is enough to make a person shit blood. That's because the intestines, which under less vigorous conditions receive up to one-third of the blood pumped with each heartbeat, can react poorly to prolonged lack of blood flow, known as ischemia. The need for the bowel to receive blood so it can parcel out absorbed glucose, salts, and nutrients to muscles in dire need of them is at odds with the demands of the muscles themselves, swollen with oxygen-rich blood, pumped straight to them in hard and fast lub-dubs for unending hours.

There is a reason that the gut is so sensitive to vasoconstriction signals. It means our bodies can rapidly shunt blood from our bellies to our muscles at a moment's notice, allowing nutritional resources and oxygen to be diverted to a necessary athletic task at hand, should the need arise. But if this diversion is sustained, and the blood flow to the gut is cut by swelling, bleeding and premature cell death can occur. Add to all that the action of shaking a bag of guts for hours on end via the mechanics of running and you, my friend, have the perfect recipe for shitting blood.

In fact, the whole digestive system can rapidly turn into an unholy and often unrestrained mess. Years ago, for an SB Nation feature on the physiological toll of sports on the body, LA-based endurance athlete and coach Jimmy Dean Freeman described ultra running to me as "an eating and drinking race, with a little exercise and scenery thrown in," and I've never forgotten it.

"If you try to stuff too much, then your body just revolts a bit and you're going to be nauseous," says Katie Wright. "You're going to be vomiting.

You're not going to keep things down. But if you're not keeping up with your nutrients, you're going to collapse and not have the energy." She says that historically in race situations, she hasn't had much luck keeping down solid food. At Big's, she's subsiding on a steady diet of potatoes, recovery drink, and energy gels made from what is essentially pureed baby food.

I've been keeping a regular log of snack sightings. There are a few minutes of activity each hour and no sleep, and I am delirious, having only slept an hour or so since the middle of the night on Saturday. To pass the time, I have been carefully recording things like the titles of songs playing, what kinds of noises the cows are making, just how drunk and miserable that one crew person was last night. And, knowing what I know about how this sport is essentially an *extreme eating and drinking contest*, a snack log.

Snacks observed: tortillas smeared with Nutella, applesauce in those little suckling pouches, bagged garlic bread, cheese slices, Pop-Tarts, dry bricks of ramen, packages of crackers like the ones they sell in vending machines, grilled cheese sandwiches (a gift offered to all runners by Laz's wife, Sandra, on the morning of the second day), single-serve prepared cups of macaroni and cheese (the powdered kind), oatmeal, grits, potato chips, sports drinks, candy bars, bottles of soda (caffeinated), potatoes, baby food, pretzels, string cheese, bananas, waffles, peanut butter. After his finish, I see Guillaume seated with an entire loaf of bread, split in twain, one half in each hand, chewing. At one point, someone takes a short drive into Murfreesboro and comes back with pizza.

The day is hard. Hard on the runners, hard to watch, absolutely exhausting, absolutely exhilarating. They say the start of the second twenty-four hours is the true start of the race, the first hundred miles a grueling initiation into the days-long suffering. The pack is thinning precipitously. When American Amelia Boone ends her race, she caps the experience by posting a close-up video of someone lancing the taut, clustered blisters that completely cover her toes. Some of the blisters are as big as cherries, others the size of deer ticks, blooming, pressed together at the surface of her cuticles like clusters of mushrooms on the forest floor. On camera, a slender, manicured hand drives a long hypodermic needle directly into the heart of the juiciest one.

Sunday is long and hot, and there are runners dropping almost every single lap. By the start of hour 35 of running (that's from 6:40 a.m. Saturday to 4:40 p.m. Sunday, for those keeping track), the field had dwindled from seventy-two to twelve. All of them make it through one of the worst laps, which is the one where the sun goes down while you're still in the woods. If you can make it through the change to darkness, and keep your pace up as night falls, then you can get back out on the road for easy ("easy"!) night laps. I feel a sense of dread as night falls; this will be their second night without sleep. They have been running so long, and some of them look like absolute garbage. Dave Proctor still looks totally fine. The fastest runners came in from the nightfall lap with four minutes to spare; the slowest, with less than one. German runner Andreas Loeffler made it into the starting corral with less than ten seconds to spare, with one of his shoes still in his hand. Later, during the night, he will fall asleep on a mailbox in the middle of the race. (What did he dream of?)

Sweden's Anna Carlsson and Tobbe Gyllebring are still in it, as are Americans Maggie Guterl, Gabe Rainwater, Gavin Woody, Will Rivera, and Shawn Webber. Shawn and his spring-loaded legs will be on crutches after the race; there will be a picture of him posted to the Big's Facebook group, propped up and smiling into the camera, a testament. Andres Villagran and his graphically injured knee are still running, the purpled inky mass swelling and spreading, obvious even in the dark. Rounding out the "dirty dozen," as Laz calls them, are Katie Wright and a fellow Kiwi, Will Hayward.

Sunday night is brutal. I am also delirious from not sleeping, miserable in the damp cold, body protesting the hard ground. Under normal circumstances, I would complain; covering this event is grueling, endless, uncomfortable, muddy, lonely. There's nowhere to shit; the toilets are full. I'm living on trail mix, dissolvable coffee granules, and a small stash of beer. I can't get warm, I can't get comfortable, my fingers have trouble taking notes in the cold night. And the funniest thing about all of this is that it's nothing, nothing, *nothing* compared to the suffering that I bear witness to each hour. I am living a dreamy, easy life compared to these demigods of pain. I feel like a townsperson watching the flagellates whip and slice their skin for penance; I could complain about my rashes and gruel, but how can one even think to gripe when there is such an astonishing display of inspirational self-inflicted misery to see?

Two hours into the second night, and the field is down to eight. Sleepy Andreas Loeffler didn't make it back in time. Gabe started the lap and came back. Anna and Will Rivera refused to enter the starting corral. A couple hours later and Andres Villagran, who injured his knee badly in hour 5 of the race, is finally finished during hour 39. Crowd favorite and popular choice for the win Gavin Woody drops before midnight, leaving only six.

Maggie, Katie, and Dave all look tough, and strong. Shawn is almost completely hobbled by shin splints. Tobbe is slowing down. And Will Hayward, a very thin, very long, very, very tired-looking man shuffles ever onward. To be honest, and I say this with a kind of stunned reverence, Will looks like absolute dogshit out there. It is hard to watch him, his slender face contorted into a motionless rictus.

"I ended up telling myself I was going to quit at the end of that second day, 'cause I was hurting," says Will, a New Zealander who now lives in Hong Kong. "I noticed that I was hurting, but I also just thought there's no chance I'm going to get through the second night. What's the point of doing four or five boring laps on the road and then just quitting at midnight? I may as well quit at, you know, six p.m. I definitely intended to quit at that point, and it was just my crew that talked me out of it." Will says that he foolishly told his crew beforehand to exercise tough love. "That, in hindsight, was a major mistake."

I ask him if he regrets telling his crew that.

"I mean, of course, no."

During the night, Will begins to hallucinate snow in the nearby fields. Shawn found his limit on lap 43, and Tobbe succumbed to his chair before starting lap 44. Katie, still running, has started her period during the night and is using what would be precious time for microsleeps to manage it. "Your mind is always going to be the first thing to give because that's what protects your body," says Katie. She then explains that a big part of running long distances involves overcoming your brain's objections to what your body is doing. "Shawn Webber . . . he pushed himself, pushed himself, pushed himself. He collapsed and then had to be carried off. I think that's very much a sign that it's the body, not necessarily the mind, that has broken down. I don't think there are many races in the world that would allow that to happen." That is, normally, your mind gives up before the body. Here at Big's, Shawn Webber pushed past that. His body broke down before his mind. It is difficult to witness.

This is a race that asks what is possible. What can the human will accomplish? How far can you really push yourself? What does it mean to find that ultimate limit? And, truly, should you?

"I feel like if you feel really great the entire time, you probably didn't run hard enough," says Maggie. And when she has to stop running? "I just feel nervous and anxious about things. I just feel kind of worthless. Not like in a deep, horrible way, but just like . . ." Her voice trails off. "You know."

The third dawn of the race is here, and the field has been whittled down to only four remaining runners: Katie Wright, Maggie Guterl, Dave Proctor, and, inexplicably, Will Hayward, who is struggling with diarrhea and dehydration. They have been running for forty-eight hours straight, a total distance of two hundred miles. Right before sunrise, two sheriff's deputies came to the race site, confused by the shuffling bodies they saw jogging up the road in the dark. Laz explained the details of the race to them, and they stayed for a whole hour, to see the spectacle. Everyone loves to watch.

Shortly after the start of lap 50, around 8:00 a.m., I see Katie walking the wrong way out of the woods. Her crew person Emma cries out and starts to run to her before she is sharply and loudly chastised by Laz, who orders her to stop and give Katie space. She does, and Katie dips under the billowing yellow caution tape that sections the trail away from the campers and begins to make her way back to the clock to turn in her ankle monitor. She is gutted. "It had come so suddenly," she says. "I hadn't really had a couple of laps of preparing myself. I didn't have that time to kind of come to terms with finishing." A leg injury has ended her time at Big's today. "There was no way I could keep that leg moving, it was very difficult to accept." Previously, her personal best for a race was one hundred miles in just under twenty-four hours. Today, she has doubled that, doing two sub-twenty-four one-hundred-mile races, back-to-back, in forty-eight hours. "It took me a really long time just to walk back to the start," she says. "I had to get down on my hands and knees to crawl down a couple of steps I had just been running over." On Facebook Laz writes:

magnificent run, miss katie. it was a privilege to be here to see it.

She takes a short nap and stays to watch the race.

With rain in the forecast, the field drops to three. Dave, who is fast and consistent. Maggie, who occasionally shows moments of duress and yet seems indefatigable. And Will, who at this point looks like a poorly reanimated corpse—clammy, pale, swaying.

At the end of the fifty-second lap, something startling happens: Dave does not come in first. In fact, Dave does not even come in second. Dave, who has been so consistent, so steady, so indomitable, comes in dangerously close to the wire, with thirty seconds to spare. Something is wrong with Dave. His crew frantically attends to him. For days, Laz has been voicing his concern that Dave is running too fast. The remaining spectators stand in hushed silence as Dave, grinning and winking Dave, Dave who smiled and ate a caramel apple at the starting line, Dave and his cowboy hat, now stands with empty eyes on unsteady legs.

"Up to this point I've never ran this well in a race, ever," Dave said on his blog. "People were telling me that I looked good, but I'm sure they say that to all the pretty girls."

But now, his vision is too blurry to read his watch, his eyes trailing his head as he looks from side to side. He's crashing. His team pumps him full of one thousand calories and a liter of hydration and puts him in the starting corral. The cowbells clang and the runners are off.

At the end of the lap, it's Maggie, it's Will, and with forty seconds to spare, it's Dave too. "Now that was a hard hour," he wrote. "All I could think was this too shall pass, this too shall pass, just get back within the hour." The roar of the crowd upon seeing him return from the woods is a jolt of motivation to him. His crew loads him into a vest filled with food and water and sends him out again, telling him not to quit. Guillaume is screaming, "Go! Go! Go!"

But Dave knew it was the end of his race. From his blog:

Everyone and I mean everyone was trying their best to urge me forward. I love these people. I entered the yard thinking again this too shall pass, get through this Dave. I didn't get far before my dragging legs caught yet another root and down I went. I got up on my hands and knees but really struggled to get to my feet. I remember crawling to a tree to use it to climb to my

feet. I stood there motionless, I'm not going to get back in time for the 53rd
bell. A lot of feelings flooded but the one that stood out was being unful-
filled. I said aloud ten times, "'I'm capable of so much more, I'm capable of
so much more, I'm capable of so much more, I'm capable of so much more,
I'm capable of so much more, I'm capable of so much more, I'm capable of so
much more, I'm capable of so much more, I'm capable of so much more, I'm
capable of so much more.'"

I walked back to the finish line. I remember it being really important
to me that I take off my cowboy hat because I was no longer racing. When
I hit the powerline clearing I broke down into tears. As if everyone was
standing there waiting for me they all stood square towards me and as if
to salute my efforts they applauded for minutes on end. I can't say enough
thank you's to all of you old friends, new friends and complete strangers.
The way you made me feel in that ugly moment of hurt meant the world
and I will never forget it.

Laz, who has been smoking cigarettes and pacing and clanging cowbells
for days gives Dave a hug, and I go into my tent and cry.

Now it is afternoon, pouring rain, and Will is quitting. He runs into
camp and makes a direct slicing motion across his throat, signaling that he
can run no more. Earlier he went into a panic when Dave dropped. He figured he was almost done and that Maggie and Dave would go on for hours,
but now without Dave in the race, it's up to him to help keep it going. When
he stops, the race is over. Maggie's race is over.

Maggie has asked Guillaume to keep Will running. "Him and Andy
Pearson, they were just giving him food and sending him back out there,"
she says. She needs him to keep running so she could keep running. And
so far, he has been. "Laz always says that once you feel like you can't win,
your race is basically over, and it's really hard to keep going," she says. She is
amazed by Will's determination to continue after hope is all but lost.

However, Will's feat of endurance and suffering and shuffling ever onward is not a solo act. His mind is gone. His continuing in the race depends
on his crew. All Will has to do is run.

"After like forty-eight hours, you don't really realize, as a runner, what's
happening outside, so you should have a very efficient crew and people who

can take care of everything you need," says Guillaume. "That's how you can go very, very far because you're not really in control any more, but people are helping you to go as far as you can." For much of today, Will's crew hasn't been sure if Will actually wanted to continue. They do not ask him how he is feeling. "We're like, okay you go back, and he was going back lap after lap, and that was amazing to witness," says Guillaume. Guillaume knows what it's like to run that far. He says everyone feels like trash.

Will sits in his chair and his crew feed him and walk him to the starting corral, positioning him in the right direction. Will is no longer quitting. The cowbell rings and Will keeps running.

"It's really easy until it gets hard, and then it's really, really hard," says Will. "But by that time, you're kind of so out of it that, in some sense, you don't even notice. The thing I noticed about that race, more than any other, is the way that time really stopped moving.

"Time was kind of endless but also instantaneous."

No one is expecting Will to make it back in time, but he does.

He does and he eats and his crew put him in the starting corral and he sways and turns to face the right direction and the cowbells clang and he stumbles after Maggie.

And then he does it again.

"I used to get to a point where I'd feel exhausted, and like, 'Ah, I have to stop now; I can't go on,'" said Will, "and then now I understand, no, no; you can go on. Maybe you need to slow down. Maybe you need to eat something. Maybe you need to walk for a while, but you can keep going, and eventually, it will get easier."

The sky has opened up. It's been pouring for hours.

Maggie returns from the sixtieth loop, bringing her total distance to 250 miles. It's Monday night at a race that started Saturday at dawn. Her headlamp is on, and it's still raining—a difficult finish to Monday's final hour in the forest before switching to the road until dawn. With dusk now giving way to full darkness, the starting corral is bathed in harsh artificial light, as it has been for the last two nights. There is a stark difference between the feeling tonight and the feeling last night. Last night, there were more runners, more activity. But tonight, as the last fingers of light fall through the trees, the act of watching the race unfold becomes nothing short of a vigil.

Our camp is quiet, the intensity of waiting for Will each lap holding all of us in fixed attention. I don't say anything about what is happening, and I do not hear it discussed around me as we all scan the black, wet woods for Will, for a beam of light trundling through the trees, as seemingly unlikely as it is horrifying and inspiring. He often returns close to the time limit. He might still be coming.

It is easy to see this as inspiration, the indomitable human spirit piloting its meat suit to the very edges of possibility. But there is horror here too, the visceral reality of a body at war with itself. You can feel the fatigue radiating off each runner at the end of their race like soft heat from a furnace. The first time I saw someone fully hit their limit, their eyes blank and body slick with exertion, I went into my tent and burst into tears. It's overwhelming to see it up close. The unfathomable made real, with all its attendant heaving and grimacing and effort. It's beautiful and terrifying and, my god, don't we love to look?

The race clock creeps toward 60:00:00. The closer it gets, the slimmer the chances of Will's return, the surer Maggie's victory. When there is no sign of Will with two minutes to go, butterflies creep back into my stomach, mirroring the way I felt when I arrived at Big's, when the race first started, the first night, the first dawn, when Will tried to quit hours ago during a rainy afternoon drizzle. I am moved by being here, by witnessing the outer edges of human potential laid bare at such close range that I can smell it. The suffering on display is disorientating and magnanimous, inviting you to look at it, offering itself to be seen in a way most pain cannot be witnessed. Here, at the race, all are allowed to stare pain in the face. Suffering is often outside of our control and, because of important things like empathy and social decorum and mirror neurons, it is necessarily frowned upon to stare at those in the vice grip of deep pain. To look at it up close, to consider it, to gasp and be moved by it. Here, the straightforward act of choosing to keep running for as long as possible gives its audience the gift of voyeurism, a chance to worship, something to recoil from and be moved by. What can a human body do, really? What can a person decide to do and accomplish? *What could I do if I tried?* And, just as importantly, *Where should I draw the line?*

Maggie stands in the gravel starting corral, her light jacket damp with rain, and we all watch in unison as the clock runs out on Will's return.

The crowd starts fucking screaming.

At sixty hours and 250 miles covered, Maggie Guterl becomes the first woman ever to win Big's Backyard Ultra, and she is illuminated by a flurry of photographers capturing her easy stance, muddy legs, a race bib that says, as if a taunt to her personally, *Last Man Standing*. She smiles as the sparse remaining crew and media scream and holler with joy. She did it. Tears warm my cheeks in the cold rain. I am watching the impossible become history.

"And then it just counts down to one, and there's no Will," says Maggie, "and everyone's confused on whether I run another lap. But because I had finished that lap and he hadn't, that was my 'one more lap.'" The winner must complete one more lap than their competitor to be declared the winner. If Will had finished the lap and then refused to enter the starting corral, Maggie would have to run one more yard alone. "Everyone had to kind of switch their brains over all of the sudden, and it was super anticlimactic."

After this herculean effort, after sixty hours of running and slipping her way through the backwoods of Tennessee, after besting an entire field of elite competitors, after becoming the first woman to win this prestigious race, Maggie is disappointed. "I was super ready to run."

As she stands newly crowned before a crowd electric with praise and adoration, a second thought begins to buzz around her. Where's Will? He hasn't come out of the woods.

Will had been feeling better when he went out on what would be his last lap. He'd slowly been making better time since he tried to quit so many hours ago, and he was doing better at eating and drinking; his antidiarrheal medicine kicked in. But then the sun went down.

"At some point I just fell asleep walking, and that's when I completely disassociated," said Will. "I suddenly had the experience that I'm back in Hong Kong, I'm walking around some old villages. I felt like I could see the houses, or I couldn't quite see them. It was very dreamlike, you know. I know they're there. Maybe I can't quite see them. I felt like there were people around, but I couldn't quite see it. They were like, in the house or behind the shed or this kind of thing.

"I was following a path. . . . I wasn't really conscious of actually following it, but just kind of being on it, maybe going back and forwards. And I felt like I was trying to find a way back to the city, but it was all dark, so I couldn't find the way, so I was like, 'Oh, I better find my way down.'"

Eventually, through his hallucination, Will begins to feel the cold. The yellow tape demarcating the trail penetrates the dream, but doesn't steal him out of it. He wanders through the village in his mind, looking for an exit, finding none. Then he sees a headlamp.

"Who's there?" he calls into the night.

"Guillaume!" the voice calls back. "Hey, Will, how are you? Is everything okay?"

"Hey, Guillaume, how did you find me so far in Hong Kong?"

"No, Will, you're not in Hong Kong. You are at Big's." Slowly, Will's mind returns to Tennessee.

"I was like, 'What are you doing here?' I'm not sure what he said, but then suddenly I was like, 'I've got to wake up.' . . . I remember kind of saying to him, 'Oh, what happened?' and he was like, 'Oh, the race is over. Maggie won,' and I was like, 'Oh, I'm so sorry.' He was like, 'Dude, you ran like sixty hours!'"

Will receives a hero's welcome back at the starting line. We scream and shout with glee and relief when he emerges from the woods. He is safe, he is fine. He did it. Will is okay.

"I never thought I was going to beat Maggie," said Will. He wasn't disappointed that he didn't win, and felt like he did very well. "Pretty satisfied," he mused. The satisfaction comes from truly running until he could run no longer, his body doing the bidding of a determined psyche for days on end. Will ran for as long as he could, touched his own limitations with his rain-soaked flesh, and stumbled back to earth to raucous praise. And then the race is over.

"You stop the race, and then your brain kind of frees up, and all of the sudden all of the pain flows back."

Dear Laz,

I am deeply saddened by your email and your characterization of my work.

My book, "Hurts So Good: The Science and Culture of Pain on Purpose," is an exploration of the myriad ways humans choose to deliberately engage in pain. It is not, as you put it, about "sadomasochism." I will note that I do include certain types of sexual masochism in my broad scope of inquiry, but it is not the point of the book, nor the main focus. My project is much bigger than such myopic, outdated ideas about pain on purpose.

The fact is, humans have been directly engaging with pain for all of recorded history. In my book there are religious flagellants and Muay Thai fighters; polar bear plunge participants and ballerinas; hot pepper lovers, circus performers, sacred rituals, and ultrarunners. There are interviews with neuroscientists, psychologists, geneticists, biologists, all manner of experts.

Pushing our bodies can be a glorious thing. My work explores the reasons why people use pain, and all the benefits and challenges we find along the way. I am in awe of the humans who display such mastery. Our bodies are capable of greatness and transcendence at levels most never approach, and the idea of seeing it up close thrilled me. I am a former athlete and dancer, with deep respect for the mental and physical toughness of endurance athletes.

You can imagine my utter elation when, back in April, you granted me a press credential for this weekend's race. I immediately began planning my trip in earnest. I cannot imagine a more incredible feat of human endurance and sovereignty over pain than your race. I am Southern and in love with the ecology of this part of the country; to get to bear witness to such an extreme display of human possibility, while surrounded by such beauty, was everything I could dream of.

So I want to be clear: Are you revoking my access to the race this weekend?

Thank you for your consideration,

Leigh

In the end, he was convinced. I was allowed to come to the race.
But as for Laz's assertion?

like many sport there is discomfort involved,
but it is a cost of competition,
not an objective.

I beg to differ.

Chapter Nine

SERIOUS PLAYTIME

Accorning to the International Association for the Study of Pain (IASP), the official definition of pain is "an unpleasant sensory experience associated with, or resembling that associated with actual or potential tissue damage." They add that, importantly, pain and nociception are different, that people learn about the concept of pain through their life experiences, and that pain is *always* (emphasis mine) a personal experience, one that is influenced by biological, psychological, and social factors. Plainly: pain isn't simply nerves firing. Everyone learns about it differently, and context is everything. It's a fresh definition too, having been updated in July of 2020, mere months prior to the writing of this chapter. Just think—the meaning of something so universally human is still being debated by the specialists who study it. Far from the tautological definition of pain as "something that hurts," the IASP definition attempts to capture the murkiness inherent in understanding and demarcating pain while still encompassing all of the wildly diverse experiences of it. I like the definition, mostly. It's wiggly, expansive. It makes room for all the ways I've felt pain before, be it intended or otherwise, and it rings true in its careful, inclusive considerations. They don't say pain is bad, which I appreciate, because it isn't that simple. Pain cannot be pegged as the opposite of pleasure. The two are not a polar binary. More like competitors, coconspirators, amplifiers, dampeners. The researchers don't say that pain is caused by harm, because sometimes it isn't. And they don't say that harm always causes pain, because sometimes it doesn't. I like their crisp definition of something so messy. It makes me think of some of my favorite questions to ponder. What does it mean to have a painful experience on purpose? And why do so many people do it for fun?

After everything, I don't actually have this fully figured out. I believe that anyone who tells you that they do is, at best, overconfident and a narrow thinker. The reason I don't have a tidy answer is because there isn't one. Not yet, anyway. This book is not a prescriptive text. I do not intend to be rigorously instructive regarding your life, your body, your choices, or your feelings. All I'm saying is this: once I noticed the propensity and enjoyment of certain kinds of pain in myself, I noticed it everywhere.

Humans have been dabbling in pain for about as long as we've had records. I am not asking you to judge that. I am merely suggesting, by way of example, that it can be illuminating to cultivate a curious eye toward your own relationship with pain. And if one is curious about pain, who better to talk to than a professional in the realm of BDSM?

"The world of sadomasochism is vast, and it represents an incredible opportunity to look at all those buttons you've been told never to push, and then to consider how to ethically and responsibly push them," writes Professional Dominant (and Dom/sub Trainer) Creature KPW over email. "Within the bounds of consent, negotiation, skill mastery, and an understanding of safety, responsible kinksters can learn to make the reality of button-pushing accessible in a positive framework within their relationships." Over two decades of being a sexuality educator, they have witnessed again and again the power of kink to expand understanding and deepen connection between partners, "although it's not always an eloquent or positive process for everyone, miles may radically vary!"

Creature is one of those just, almost upsettingly, luminous people: all impish and swaggering and charismatic and present. It's too much! I personally have witnessed much melting in their presence. And if there is a brain I want to pick on this, it's theirs. A sexuality/gender/identity educator, BDSM skills and safety instructor, and performance artist, Creature has logged a significant amount of time exploring pain for fun. So, in their professional opinion, what's the draw?

"One of the most shocking and empowering things I learned when I first started down the road of exploring my own desire for masochistic experience was that my body, my mind, and my heart proved far stronger than I had been taught to believe," says Creature. They said that as a transgender individual raised female who came out as nonbinary later in life, their experiences within BDSM helped them learn to understand themselves better, and gave them a safer way to do that. Learning to assimilate "asked for" pain into their life taught Creature exactly how much emotional, psychological, spiritual, and identity-centered pain they were already experiencing nonconsensually in their daily life, which began to reframe everything. Their experiences in BDSM pointed to a certain inner toughness that had perhaps not been immediately obvious. "I began to see more clearly what layers of resiliency I was already accustomed [to], which made it easier to locate more natural boundaries over what I had been trained to accept."

The novel concepts of play and sensation in BDSM brought clarity to them, just as it has to me and so many others. "I began to see the shape of my own psyche and my potential more clearly than I had in all of my years prior to taking this 'perverted' plunge into depravity." It's liberating to get to know yourself empirically rather than base such intimate ideas of selfhood off of expected social standards. For Creature, and for me and so many others, BDSM provides a space to play and to learn.

But what is it like to watch the effects of their sadistic expertise? I have always known Creature to be a scrutinizing observer.

Sometimes pain processing is very loud, they tell me. Or it can be silent. There can be a physical release, maybe a deep letting-go, which can sometimes look euphoric, or even be filled with laughter. Many people ride a wave of alternating reactions, tensing up and relaxing as the painful stimuli come and go. There are people who disassociate, or go hypervigilant. Some plead for it to end, using everything but their safe word; others might take a stand, egging on the pain-giver with bratty taunts. "Sometimes processing sensations looks a lot like a roller coaster of intense emotions streaming out of a person's face and body, or they will become eloquent, heady, or even stoic." Sometimes it can look quiet, like an escape into the dazed, adrenaline-soaked world of subspace. "There are many ways to enjoy, process, survive, and welcome pain into one's lived experience, and just as many reasons to bring the exploration of this concept (and reality) into your relationship or life." There is no one way to do this, just like there is no one way to feel pain. I love the endlessly diverse ways people experience masochism, and I think nearly all of those descriptions could apply to a kickboxer during a grueling workout or someone following Wim Hof up the side of a frozen mountainside with no shirt on. I think it really captures something true and mutable about the human experience with deliberate pain, and I will be forever curious about all the reasons why people do it.

BDSM is, of course, not new. For example, prior to industrialization and urbanization in the West in the 1800s, BDSM was either the province of religious and cultural movements (like Lupercalia or the religious flagellants) or wealthy elites who had the privacy and means to indulge in their fantasies. Surely other Westerners had similar desires and fantasies, but these would have been more difficult to arrange in a consensual context. Large extended families and grueling workloads would not have lent themselves to pleasurable pursuits. Also, there are scant records of the lives of commoners during this time, so it seems that most nonreligious and nonroyal acts of BDSM during the Western medieval era have been lost to history.

However. Throughout ancient and medieval history, corporal punishment was ubiquitous. Scenes of physical violence against prisoners, serfs, and livestock would have been common for most people. The whip was a widely recognized symbol of power. Later, caning of children in schools and employees in shops contributed to this milieu. Divisions of power were stark and in plain view of the public. So it wasn't much of a psychological leap for people to fetishize these kinds of acts.

The late eighteenth century in England (and Holland) and the nineteenth century in much of continental Europe saw the rise of capitalism, urbanization, and disruption of conventional extended family units. People (mainly men) moved around, lived in large, anonymous cities, and weren't expected to marry and begin raising children as early. People (mainly men) had disposable incomes for the first time, especially among the newly emerging middle and bourgeois classes. Newspapers began running personal ads for people to seek out like-minded people. Queer and kinky people could find each other, and roughly one hundred years later, the internet made it even easier. The activities associated with modern BDSM have been around for thousands of years, but that still doesn't answer the question of *why*.

Scientists are looking into this. Research on BDSM—the alphabet soup acronym for bondage/discipline, dominance/submission, and sadism/masochism—is still somewhat limited, though there are signs the tide is turning. There's been a recent spate of new work on the topic, much of which was used for this chapter, notably a 2019 survey of current BDSM literature, "Physical Pain as Pleasure: A Theoretical Perspective." Published in the *Journal of Sex* and rigorously undertaken at the University of British Columbia, the paper takes a big, wide look at all the things researchers are finding out about why people get into pain for fun. The researchers looked at reasons why people desire pain in the context of consensual BDSM (which, I should note, is a redundant phrase: if it's not consensual, it's not BDSM—it's abuse). They found that, broadly, many reported reasons were surrounding submission, contrast, and achievement. And in my own interviews with self-identified masochists and people who experiment with pain on purpose, these trends were apparent there as well. Below are just a few of the common responses I've received over the years, responses to my question of *why* people do this and *what* they get out of it.

A very popular reason for incorporating pain into sex is that it can enhance feelings of helplessness and submission, which can intensify power play. Receiving and tolerating pain can feel like giving a gift to the person causing said pain, turning endurance into a test of devotion. Giving up control is just as much of a powerful aphrodisiac as wielding it, and pain certainly provides a fast track to finding and testing such limits.

As James* told me, "The only times I have cried as an adult have come from being flogged to the point of safewording. For me that is a function of trust." Pain as a pathway to submission as a pathway to catharsis.

Another popular reason for desiring pain is that it creates a contrasting sensation that intensifies pleasure. All pleasure all the time sure sounds nice, but a little pain mixed in can really boost the experience of both, the way people add salt to syrupy-sweet caramel. For many people, it's exciting to

---

* Name anonymized to protect this person's privacy.

be on edge like that, touch brimming with possibility. That any one gentle neck kiss could turn into a bite keeps things exciting, like having a little safe danger on the side.

Other masochists and pain players cite as the main draw the appeal inherent in the challenge of enduring pain, seeking to create feelings of achievement through their play. A sense of pride can be cultivated through sustaining an ordeal, even if it is self-inflicted. I'm certainly extra blissed-out and woozy after a scene if I feel like I took more pain than usual, with all of my ballet programming lighting up like a dusty switchboard. For those who carry cultural messages of suffering as worth, be they saint or ballerina or pepperhead, pain on purpose can cause a delicious sense of accomplishment. As one person told me, for them it's about testing limits, which builds to surrender and release, a journey they find deeply fulfilling.

But that's just a few of the reasons that came up in the studies I surveyed. There are so, so many other reasons for desiring pain. For example, lots of times, people just want to check out, get some distance between them and their thoughts. Other times, these same people might do the opposite, trying to explicitly work through the feelings they often so dearly need space from. (I am both of these people.) People tell me of pain as a way to discharge energy that feels trapped, of pain as a way to create something to focus on outside of their thoughts, of pain as way to conjure up some feel-good brain drugs. (Yes, please.)

Some people say they like the more formalized structure of negotiated scenes, which helps mitigate their anxiety and allows them to be less worried and more present in the moment. Others are in it for the taboo. Some are in it to chase a quiet brain. Some people say that pain just feels good to them. Others, that the unpleasantness of the pain is worth the payoff of the high that follows. Some do it to feel close to their partner. Or to know themselves better. Or to just try new things. Or just because it's fun to them. All of these reasons are overlapping. Like I said, it's delightfully messy.

Perhaps for me the thrilling exploration of consensual pain is a reclamation. A rewiring of sorts. I have endured a lot of pain for a lot of different reasons in my life, so maybe I'm just trying to rewrite my stories with different outcomes. Maybe I want to suffer and hear that I'm good at it, gifted, exceptional, instead of having some small, sweaty British man scream at how pathetic and disgusting my soft body is. Maybe I want someone I love to say perverted shit about my soft body while they fist-fuck me and spit in my face and make me come all over the floor, the new memory of these words running slick all over my thighs while I giggle in the pinkness of bliss. Maybe instead of feeling incapacitated by my corporeal form, I want someone to beat my body senseless so that I can feel powerful, irrepressible, resilient, sturdy in the face of fear and hurt and pain. Maybe nature, nurture, and ballet plainly conspired to teach my brain to love this shit. Maybe it's all of these things and many others that I haven't discovered yet. Maybe it's something else entirely.

I mean, we can analyze this shit all day, and come up with endless theories, but regardless, these two things are true: I can intellectualize the origins of my desires, exploring their potential root causes and viewing them through an analytical lens; and my pussy gets wet when it gets wet and I like what I like.

During a scene with my partner, with whom I have had years of negotiated, violent sexual encounters, I'm not thinking theory. I'm not wondering why I'm like this with some flinty remove; I'm not thinking, "Hello, it is trauma-processing time." I'm not doing that. I'm getting punched in the face and then having several orgasms in a row. I'm getting Saran-Wrapped to another person and hung from the ceiling and beaten hard until I stink of arousal and fear. I'm collapsing in a puddle of tears and piss and blood and come and smiling like a vacant Labrador and then I'm wandering back into the house and having a little snack and taking a shower and snuggling up for a cuddle and then, maybe after all that, with an ass so sore that I cannot sit right, I might think to myself: *Hmmmm. I wonder how I turned out this way?*

In writing this, I wanted to understand why I am like this, why any of us are like this. And the answer I've uncovered, years into detailed research, countless interviews, dozens and dozens of books and records, hundreds of papers, textbooks, manuals, letters, hushed conversations in the corners of bars, and interviews conducted over late-night emails and endless social media platforms? The answer is this: I'm still not sure, but there sure are a lot of us.

The reasons why people get into pain seem as endless and mutable as humans are themselves, true. But, at the end of the day, they all fundamentally have one thing in common: trust and consent. Pain play requires consent, period. Without consent, pain isn't yours to control. No consent, no pleasure. Receiving pain you don't consent to is abuse. And the people who abuse others under the guise of BDSM perpetuate deep harm.

I experienced this firsthand. Years ago, I was abused under the guise of BDSM, which made me uncertain how to escape it. But I am safe and free now, and the renewed trust and safety I have found through consensual play is a source of great joy and light in my life.

Because that's the thing about consent. Consent, in any realm, is what makes pleasure possible, be it having a bite of a nice meal or receiving a warm hug or having an orgasm.

You have to decide to engage with a thing in order to receive pleasure from it. Context for pleasure is just as influential as context for pain. A lover may touch me where a stranger may not; a lover may pull my hair so hard I cry, but a stranger will be curtly reprimanded for merely an overly familiar arm touch. Context, boundaries, consent. Context, boundaries, consent. The pain I seek depends on context, boundaries, and consent. Otherwise, I'm just suffering.

"A Dom who is selfish and causes pain that is uninvited, or who is careless about follow-up from their play, can perpetrate real harm," says Creature. But one who is thoughtful, caring, with tremendous respect for boundaries? Who can take responsibility for negotiation and check-ins and aftercare? That's the kind of person I want to play with. Creature says that compassionate BDSM tops are capable of helping to expand their partner's resiliency and bandwidth in many aspects of their life, but they are careful to note that BDSM is *not* therapy. They are also quick to state the necessity of trust.

Creature adores helping people identify and expand their boundaries, an engagement that can only happen with deep wells of trust.

"I must take responsibility for my actions: intended fallout or not. I must be vulnerable and brutally honest. I must listen to the needs of my partners." These are the ways trust is built. And with trust, so much more is possible.

That's the thing about trust and consent. Once trust is established between two or more people, and assuming they continue to choose to preserve and respect those sacred tenets of engagement, a great depth of connection becomes available to them. Suddenly, wild, bananas-fun shit is on the table, like exploring boundaries and fantasies, feeling seen for who you really are, testing out varying degrees of vulnerability, getting paddled until you radiate the vermilion, you get the point. People ask how I can play so hard with certain people and yet will sometimes refuse even a perfunctory party spanking from others and the answer, for me, is trust. And, a little bit, sometimes, love.

Creature agrees. "I've certainly had the experience of asking someone to flog me, and then (surprisingly) found myself hating the sensation because the person I asked didn't take the time to connect with me energetically before they laid down a strike on my body." On the contrary, they've thoroughly enjoyed suffering through the far more painful sensations garnered at the business end of a bullwhip, if they feel trust, love, and connection with the person holding it.

The vulnerability that can come with this kind of play is a major reason why conversations that happen before, during, and after a scene are so important. Negotiations, check-ins, and aftercare are basic expectations of BDSM because players are agreeing to explore delicate, provocative pockets of existence, and care must be taken to keep everyone safe.

"The brain is the largest pleasure-inducing and trauma-recording organ we have, and as such, momentary perceptions of the mind will affect what pleasure and what pain is even available to our bodies at any moment in time." Once you accept this perspective, Creature says, it becomes easy to see how the very ideals and practices of negotiation, consent, check-ins, and aftercare are direct tools for creating positive sensations and outcomes for pain players.

With trust, we are able to process and enjoy far more experiences and sensations than we could without it.

Trust and consent together, then, conspire to transform pain into a territory that can be intentionally explored if one simply has the inclination and curiosity.

But where does that inclination come from? We are each the products of our culture, our trauma, and the experiences in our lives that have shaped our deepest wants and desires. I like all sorts of painful shit and actively court high-sensation, risky-feeling situations. Someone might say that is a product of my abusive upbringing in ballet and the internalized messaging

I was exposed to as a child about pain and worth and discipline and what makes a person deserving of love. I am sure there is some truth in that, but it is not the only truth. I think there is something more universal to be noticed in the way we use our bodies for fun, and how pleasure is more complicated than simply *This feels good.*

These days, I like what I like. I get my needs met in ways that are consensual and not harmful to me. I am the product of my life experiences, trauma, healing, history, relationships, family—all of it culminating in my own personal blend of desires. It is what it is. This is not the way everyone comes to play with pain, but it's how I got here, and that's okay.

It's important to note that multiple studies have found that rates of mental illnesses are not higher in BDSM practitioners as compared to those of the general population. Trauma isn't a prerequisite for getting into this kind of thing. The physiological response to noxious stimuli is a plaything available to most humans who want it, and the interpersonal bonding facilitated by shared painful experiences is not the sole province of those who need it most. But it's also impossible to delineate "physical" and "emotional" responses to pain as separate from the circumstances that influence our perception of it. Expectation, mood, environment—all of these things shape our experience of pain, which the brain creates fresh for us, anew each time.

So, yes, we are all wired similarly enough, but why do some people seek out more intense pain than others? Why do some people wade deep into pathological territory, up to and including the point of death, while others go about their lifetimes with little more masochism than nail biting, Tabasco sauce, and light erotic spanking? I think there is something to the idea that both pain and pleasure exist on a spectrum of desire and expectation, and that wanting something is as important to the ultimate experience as the nature of the stimuli itself.

In that way, it is what we are expecting and what we desire that make possible the pleasure we seek, even if that pleasure hurts. Consenting to pain means you have the power to stop it, which puts it all on your terms. There is power in that.

In some ways, it was easier for me to reveal myself to you at the beginning of this book. Confessing to or even just speaking of sexual proclivities can

be a little titillating, sure, but revealing myself after a fire hose of vulnerability hits a little different. Shocking people is easy. Allowing yourself to be
seen and possibly even understood? Terror. It's easy to tell you stories that
have endings, but it's much harder to allow a life-in-progress to be witnessed.
I don't know how the story ends. I just know that for now, I feel happy. And
pain plays a part in that joy.

Chapter Ten

BLISS

IT'S September 13, 2020. My slide into seasonal depression pro-
ceeds apace, but pales so frankly in comparison to the daily specter of the
news that it barely registers. A year ago today, I was flying across the country to eat the world's hottest pepper in a dusty fairground parking lot. Six
months ago was the last time I'd been in public. These days, I'm sequestered
at home with the luxury of a remote job during a global pandemic.

On this day, my own little Saturday unfolded in a little pocket of bliss, a
dissonant feeling, precious, somewhat nauseating and unsettled. The West
Coast is burning in an unprecedented fashion, and social media is flooded
with pictures of red skies, pleas to help find loved ones, and DIY hacks
for mounting filters to box fans. Enormous storms are being born on hot
oceans, and Sally is busy becoming a hurricane that is preparing to strike
the Gulf Coast in two days. The pandemic is still sweeping through my city,
which is incongruously filled with tourists and hoteliers. The election looms.
But I am okay. Healthy, fed, housed, in love. It is hard to square the horror
of what is happening with the quiet joys of home. It is hard to write about
privilege and luck and joy in the face of white supremacy and police bru-
tality and climate change and novel coronaviruses and political corruption
and corporate greed and hunger and dirty water and all the suffering that is

choking my country and my planet. The air stinks with grief and fear. These are frightening times.

The room is set up with a big red sheet draped over the bed and a small table. I have been thinking about this for weeks, for months, for years. When was the last time you were truly scared by something that you desperately wanted? It's all I could think about over breakfast this morning, eating buns filled with eggplant and honeynut squash from the garden, sprinkled with chopped chili peppers, my mouth burning while Casey and I hunched over word puzzles and coffee. It's all I could think about when I took my afternoon run, my brain frantic as usual, electric fantasies bubbling up through the soup of work and worry. I ran laps around the house for an hour, making tight circles on a worn path like some kind of agoraphobic monk, passing Casey over and over as they meticulously checked a lumber order. I tend to run a little faster when they're outside, lengthening my stride just a little bit, landing a little lighter on my feet, wanting to impress. I am deep inside a daydream, lungs heaving, mind pushing body. It feels like I've been thinking about tonight forever, and I thrill at the thought of bringing to bear a private desire that has taken years to gather the power to explore.

In the room, they tell me to rinse off, and I do. They ask how I feel, which is nervous. I ask them if I should wear anything, my emotions already careening into subspace, suddenly feeling shy and exposed. They say no, quietly, as they set out two different kinds of disinfectant, a box of gloves, a box of gauze pads, a sharps container, and a box of needles. I shiver, just a little. Casey is lithe and captivating, with the ability to focus their attention on me in such a way that makes me fumbling, the color of cured ham.

Often our scenes are chaotic, with heavy hitting and improvised implements, but today is going to be more formal, a celebration. My hands are cold and clammy with nerves, but my bare chest feels warm with love and desire and swooning. I feel absurd, all cartoon heart-eyes and fast breath. It's hard to explain to people that my most violent (and beloved) sexual encounters are coming from the most kind and loving partner I've ever had. Sometimes I worry that people might think that I am bad for wanting it, or they are bad for enjoying it, but you've read the book now, so maybe you'll understand.

They instruct me to lie on the bed, prone. They gently sterilize my back, my bottom, my thighs, kneeling over me with such care I feel like my sternum might pop from the affection. I struggle to stay still, nervous energy surging through my body. I gingerly lift my chest from the bed like an amorous seal and take a sip of water from the glass they've set out for me. I wiggle my toes. I listen to the sounds of preparation as my body shines under lamplight, and I try to get my mind right, thoughts racing with the mechanics of nociception, sucking in deep lungfuls of air to stanch the fear that's priming me for pain. Of course I want it to hurt, but I want it to be manageable in such a way that I can take a lot of it. The truth is, after six years of playing together in this way, Casey is very good at priming me for pain. They know what scares me, how to incapacitate me with fear, how to get me sobbing before the first blows even land. Today, though, the ritual is different. The fear is not from body position or something over my head or balancing painfully atop a table, body swaying in mean ropes. No, today the fear is the antiseptic smell of a doctor's office, the plastic crinkling of medical supplies being opened.

Are you ready, they ask, voice low. I say yes, which is true, sort of, or at least as true as it can be. I am terrified of needles, and my body is positively thrumming in wait. Senses heightened by the knowledge of what's to come, I hear them pop the cap off a hypodermic needle. *Take a deep breath, baby*, they say, and I do. *I'm so proud of you.* And just like that, I start melting through the bed.

I don't know where I am going to be pierced today, or how many times, or what it will feel like. I have had a couple cartilage piercings in the past, the sketchy kind that you get from the guy who washes dishes at the local sushi restaurant, when he is off work and drunk in your friend's kitchen. I have many hours of sitting still for tattoos. I also have years of bad medical experiences, being in and out of hospitals, almost dying, getting improper care from overzealous sociopathic med students.

But this is play piercing. This is different. I'm in charge.

The first long, slender needle goes in, slowly. It feels like it's precariously close to the inner parts of my buttcrack, but when I look at pictures afterward, I see that it is about four centimeters away from where I thought it was, more inland on the expanse of my muscle butt. It feels so sharp, which is a stupid thing to say because it is so sharp, but my loud brain could sound no other alarm. Casey pushes the needle through me, slipping it under my skin and then out again, like they're pinning a hem to be stitched. My brain sets off some desperate visuals, and I see the color orange (but only inside my head, if that makes sense). There is a steady *nonononono* rising from my skin, a shrill request to stop tempered by my desire not to. I take deep breaths and try to remain relaxed and calm. I know what worry does to pain. How funny is this game that I love so dearly, to hurt and search my mind for ways to avoid it, all the while asking for more pain. Resiliency as fetish.

Before we started, Casey instructed me to tell them if I felt nauseated or faint. Sometimes people pass out, throw up, have a bad time. Which, I mean, of course. Bodies are fundamentally self-protective, a simple fact that makes masochism feel so delightfully naughty. I love feeling like a good boy, still under this needle. I love feeling a little bad, like I shouldn't be doing this. The discord between my desires is delicious. They pierce my ass again, a second needle going in slightly below the first, slowly, slowly, way too slow. I feel a flash of light, but it's just the electricity of my body screaming a most practiced aria.

It is easier for me to admit feats of strength and traumatic experiences than it is to show you the squishy underbelly of how tender I really am. It's easier to describe taking my tampon out next to a mountain of shit in a Tennessee porta john, or retching in a California parking lot, than to admit that I am soft, happy, and in love. Maybe there is a theme here. Maybe it is also easier for me to open myself to love if there's a sharpness inside too. Like how I am too ticklish to be touched lightly by most people, in most places, but roughness always feels more manageable, easy to enjoy. A gateway to intimacy, perhaps. But, really, who knows? Nature, nurture, genes, parents, experiences and hopes and tragedies. Life unfolds as it does, and I am here to feel it surge through me in all of its curious ways. I'm not sure that there is one answer to why I am like this, to why any of us are; I am sure that there are millions of answers.

Casey slides the second needle in about a centimeter below the first, then a third. They're working so slowly, or maybe time has slowed down for me, same thing from my vantage. Mostly, they handle my body in silence, and each directive from them to breathe feels like a lifeline, a tether to them, my body, this life. They put their mouth next to my ear and tells me they love me and hot tears leak down my face. Needles four, five, and six go into my other cheek, and they tell me that they're giving me a little break to adjust, brushing their fingertips along my pierced softness. They kiss me and it's drugs, just absolutely drugs. Their soft mouth, their hot skin, my soft mouth, my soft skin, everything drugs. Without moving a muscle, I tip forward into the great expanse of loving, of sensation, of *pain*. Eyes closed, I feel like I am falling. In the darkness behind my eyelids, I know, resolutely, of the love in that room, as sure of it as I am of the mattress beneath my body, of the windows letting in filtered late summer light, of the needles in the skin.

Are you ready for more? they ask, kissing the top of my head and smoothing my hair. I wiggle my toes and say yes, bringing my focus back to my breath. The next needle goes into the top of the back of my thigh and my brain bucks like a bull. It's horrible, this secret place on my ticklish body, unaccustomed to such attentions. But they tell me to breathe and I do it. They put another needle in, and another. They keep going. They put three more needles into the back of my other thigh. My brain cannot parse the pattern. To me, it feels like each needle is going in on top of the other, re-piercing the skin from different directions, pulling the layers of me apart like pastry. But that's just my brain making do with its usual best guess: the six needles in my thighs are in fact arranged neatly, three in a row, each a couple centimeters apart, just like the other six.

We take another break. They kiss me again, again, again, and I kiss them back with unrestrained affection, urgently pressing my mouth to theirs, as if I kissed them just right, they'd know how happy they make me. I jiggle my ass and feel the metal inside me, the way my skin is tugged at by the intrusions, the electric sparks of nerves sounding the alarm. They rub my piercings with their gloved hands and my stomach does a sickening lurch, butterflies competing with fear competing with joy. I am delirious with all of it.

We do two more piercings, this time in my upper back. The second one hits something awful and I cry out in desperation. Up to this point, the piercings are hard to bear, but I had moved through them by breathing big and slow, by letting the muscles of my body contract in great, heaving ripples. But that last one, needle number 14, which meant holes 27 and 28, respectively, that one does me dirty. It bleeds just a little more than the others. It makes me sob in big, crashing waves. I do not know if I am really crying on the outside or what my face is doing or what sounds slip from me in my stripped state. I don't know what headlines are rolling in, what day it is, anything. I just know that there are twenty-eight holes in my body, a gift from someone who I love with all my slick, wet cells, my deep red marrow, every gooey extrusion that seeps from my fragrant body. I know they love this too, love me, want this. In that moment I am a hollow shell of a skull, knowing almost nothing, feeling deeply the love in the room, the pain in my skin, the bed pressing up against my body, their hands. They ask if I want to be photographed, and I do. I want to do anything I can to capture this ephemeral bliss. I feel like I have taken MDMA, and all I want to do is taste the inside of their mouth, rub their tongue along mine, pitch headlong into the abyss of adoration.

I feel safe and loved. I am safe and loved.

Each time they pull a needle out of me, I sob into the sheets. Isopropyl burns me from the inside out, and these last fourteen moments of pain are the most exquisite and maddening and scorching moments of this scene. The stinging of antiseptic lasts much too long and is over immediately. My whole body erupts in a surge of euphoria and I love them, I love them, I love them. Every single scene with every single person is different every single time, but this one, this one has so gently flayed my heart and sewn it back inside me with such softness that I feel the thrum of the raw capacity of my body in a way that is animal, rare. We wet the sheets with sex and I collapse into them, happy.

I float back down to earth with all the urgency of a dandelion seed, dreamy and unburdened by the pull of gravity. I pad to the bathroom, breaking the spell slowly, gingerly hovering over the toilet seat so as to not sit on my freshly punctured skin. They rise, and I let them hold me upright, face pressed into their chest, knees barely functioning. "Oh my god, I cannot wait to do that to you," I say, drooling the words.

"Yes please," they say. "Anytime."

ACKNOWLEDGMENTS

THROUGH THE PROCESS OF WRITING THIS BOOK, I HAVE BECOME A REPOSI-
tory for hundreds of precious stories, the keeper of secrets entrusted to me by
masochists and pain enthusiasts the world over. This book would not exist
without the people who boldly shared with me the intimate details of their
nuanced and often complex relationships with pain on purpose. For them, I
am eternally grateful.

My deepest thanks to my agent, Anna Sproul-Latimer of Neon Literary,
for her uncanny ability to make me feel like I can do hard things and her
resolute belief that I could pull this off. She has an incredible capacity for
curiosity and scrutiny, and this book was made better by all the ways she
asked me what, exactly, I was trying to do here.

To my editor Colleen Lawrie, whose willingness to jump into strange
waters with me was matched only by her insistence that we get it right. Her
deeply considered assessment of my work improved it beyond measure, and I
am so thankful for the time she spent with my book.

An enormous thank-you to Brooke Parsons, Lindsay Fradkoff, Kaitlin
Carruthers-Busser, Christina Palaia, and Pete Garceau.

I owe a debt of gratitude to the many experts who helped shape my understanding of pain, including Brock Bastian, Will Hamilton, Elizabeth
Harper, Allan House, Creature KPW, Hane Maung, Paul Rozin, Stephen
Stein, and Jens Foell, who is the most helpful researcher I have ever encountered in my life. Foell has a seemingly boundless capacity for answering my
seemingly boundless questions, the limits of which I have sought but not yet
found.

To the hot pepper eaters, thank you all for sharing your joy with me, but especially Shahina Waseem, Dustin Johnson, and Greg Foster. I grew and ate my own Carolina Reaper peppers this year, and it's all Ed Currie's fault.

Thank you to the Coney Island Polar Bear Club for its annual New Year's plunge and to Rain, who so graciously deigned to meet the chilly brine with me. Thank you to all the brave souls who walked me through the contours of their challenging memories, but especially Anna Gioseffi, Darien Crossley, Sarah London, and Dan Block. Thank you to the Big's Backyard Ultra and Gary "Lazarus Lake" Cantrell, who allowed me to witness an astonishing days-long race. Thank you to all the runners who let themselves be seen in their endeavors, especially the ones who took the time to explain what it feels like to run hundreds of miles in one go: Guillaume Calmettes, Maggie Guterl, Will Hayward, Dave Proctor, and Katie Wright.

And of course, none of this would be possible without the unrelenting support of my child and my partner and my cat. To Eliotte: you inspire me to greatness with your fierce curiosity and your tender heart. Loving you has taught me what loving can be. To Casey: your steadfast insistence on treating me right has revolutionized my entire life. When I consider the terrifying ordeal of being known, I turn bright pink with the joy that comes from being known by someone as good as you.

And to Larry Hotdogs, my office manager: I simply never could have done it without you.

BIBLIOGRAPHY

From the Top

Birnie, Kathryn A., Patrick J. McGrath, and Christine T. Chambers. "When Does Pain Matter? Acknowledging the Subjectivity of Clinical Significance." *Pain* 153, no. 12 (2012): 2311–2314. doi:10.1016/j.pain.2012.07.033.

Busch, Volker, Florian Zeman, Andreas Heckel, Felix Menne, Jens Ellrich, and Peter Eichhammer. "The Effect of Transcutaneous Vagus Nerve Stimulation on Pain Perception—an Experimental Study." *Brain Stimulation* 6, no. 2 (2013): 202–209. doi:10.1016/j.brs.2012.04.006.

Choi, Kyung-Eun, Frauke Musial, Nadine Amthor, Thomas Rampp, Felix J. Saha, Andreas Michalsen, and Gustav J. Dobos. "Isolated and Combined Effects of Electroacupuncture and Meditation in Reducing Experimentally Induced Ischemic Pain: A Pilot Study." *Evidence-Based Complementary and Alternative Medicine* 30, no. 1 (2011): 1–9. doi:10.1155/2011/950795.

Dickenson, A. H. "Editorial I: Gate Control Theory of Pain Stands Test of Time." *British Journal of Anaesthesia* 88, no. 6 (2002): 755–757. doi:10.1093/bja/88.6.755.

Floyd, Nancy. "People Issue: Ballerina-Turned-Boxer Sarah London." *Nashville Scene*, March 8, 2018. https://www.nashvillescene.com/news/cover-story/article/20995171/people-issue-ballerinaturnedboxer-sarah-london.

Garland, Eric L., Barbara Fredrickson, Ann M. Kring, David P. Johnson, Piper S. Meyer, and David L. Penn. "Upward Spirals of Positive Emotions Counter Downward Spirals of Negativity: Insights from the Broaden-and-Build Theory and Affective Neuroscience on the Treatment of Emotion Dysfunctions and Deficits in Psychopathology." *Clinical Psychology Review* 30, no. 7 (2010): 849–864. doi:10.1016/j.cpr.2010.03.002.

Glick, Robert A., and Donald I. Meyers. *Masochism: Current Psychoanalytic Perspectives*. Hillsdale, NJ: Analytic Press, 1988.

Greenberg, Jordan, and John W. Burns. "Pain Anxiety Among Chronic Pain Patients: Specific Phobia or Manifestation of Anxiety Sensitivity?"

Behaviour Research and Therapy 41, no. 2 (2003): 223–240. doi:10.1016/s0005
-7967(02)00009-8.

Hansen, George R., and Jon Streltzer. "The Psychology of Pain." *Emergency Medicine Clinics of North America* 23, no. 2 (2005): 339–348. doi:10.1016/j
.emc.2004.12.005.

Hockenbury, Don H., and Sandra E. Hockenbury. *Psychology.* New York: Worth Publishers, 2010.

Hui, Kathleen K. S., Ovidiu Marina, Jing Liu, Bruce R. Rosen, and Kenneth K. Kwong. "Acupuncture, the Limbic System, and the Anticorrelated Networks of the Brain." *Autonomic Neuroscience* 157, nos. 1–2 (2010): 81–90. doi:10.1016/j
.autneu.2010.03.022.

Iannetti, G. D., and A. Mouraux. "From the Neuromatrix to the Pain Matrix (and Back)." *Experimental Brain Research* 205, no. 1 (2010): 1–12. doi:10.1007
/s00221-010-2340-1.

Kaptchuk, Ted J., and Franklin G. Miller. "Placebo Effects in Medicine." *New England Journal of Medicine* 373, no. 1 (2015): 8–9. doi:10.1056/nejmp150
4023.

Kaufman, David Myland, Howard L. Geyer, and Mark J. Milstein. "Neurologic Aspects of Chronic Pain." In *Kaufman's Clinical Neurology for Psychiatrists*, 8th ed. Elsevier, 2017, 307–324. doi:10.1016/b978-0-323-41559-0.00014-9.

Kopf, Andreas, and Nilesh B. Patel. *Guide to Pain Management in Low Resource Settings.* Seattle: IASP, 2010.

Legrain, Valéry, Gian Domenico Iannetti, Léon Plaghki, and André Mouraux. "The Pain Matrix Reloaded." *Progress in Neurobiology* 93, no. 1 (2011): 111–124. doi:10.1016/j.pneurobio.2010.10.005.

Leung, Lawrence. "Neurophysiological Basis of Acupuncture-Induced Analgesia—an Updated Review." *Journal of Acupuncture and Meridian Studies* 5, no. 6 (2012): 261–270. doi:10.1016/j.jams.2012.07.017.

Melzack, Ronald. "Gate Control Theory." *Pain Forum* 5, no. 2 (1996): 128–138. doi:10.1016/s1082-3174(96)80050-x.

Moseley, G. Lorimer, and David S. Butler. "Fifteen Years of Explaining Pain: The Past, Present, and Future." *Journal of Pain* 16, no. 9 (2015): 807–813. doi:10.1016/j.jpain.2015.05.005.

Moseley, Lorimer. "Why Things Hurt." Filmed in Adelaide, TEDx Talk, 14:22, published November 22, 2011. https://www.youtube.com/watch?v=gwd-w LdIHjs.

Mouraux, André, Ana Diukova, Michael C. Lee, Richard G. Wise, and Gian Domenico Iannetti. "A Multisensory Investigation of the Functional Significance of the 'Pain Matrix.'" *Neuroimage* 54, no. 3 (2011): 2237–2249. doi:10.1016/j
.neuroimage.2010.09.084.

Murakami, Haruki. *What I Talk About When I Talk About Running.* New York: Vintage Books, 2009.

Neblett, Randy, Howard Cohen, YunHee Choi, Meredith M. Hartzell, Mark Williams, Tom G. Mayer, and Robert J. Gatchel. "The Central Sensitization Inventory (CSI): Establishing Clinically Significant Values for Identifying Central Sensitivity Syndromes in an Outpatient Chronic Pain Sample." *Journal of Pain* 14, no. 5 (2013): 438–445. doi:10.1016/j.jpain.2012.11.012.

Neuroskeptic. "The Myth of the Brain's Pain Matrix?" *Discover*, January 9, 2016. https://www.discovermagazine.com/mind/the-myth-of-the-brains-pain -matrix.

Vlaeyen, Johan W. S., Geert Crombez, and Liesbet Goubert. "The Psychology of Chronic Pain and Its Management." *Physical Therapy Reviews* 12, no. 3 (2007): 179–188. doi:10.1179/108331907x223001.

Wall, Patrick. *PAIN: The Science of Suffering*. New York: Columbia University Press, 2000.

Zhang, Ruixin, Lixing Lao, Ke Ren, and Brian M. Berman. "Mechanisms of Acupuncture–Electroacupuncture on Persistent Pain." *Anesthesiology* 120, no. 2 (2014): 482–503. doi:10.1097/aln.0000000000000101.

The Wet Electrics of Pain

Besson, J. M. "The Neurobiology of Pain." *The Lancet* 353, no. 9164 (1999): 1610–1615. doi:10.1016/s0140-6736(99)01313-6.

Besson, J. M., and A. Chaouch. "Peripheral and Spinal Mechanisms of Nociception." *Physiological Reviews* 67, no. 1 (1987): 67–186. doi:10.1152 /physrev.1987.67.1.67.

Biga, Lindsay, Sierra Dawson, Amy Harwell, Robin Hopkins, Joel Kaufmann, Mike LeMaster, Philip Matern, Katie Morrison-Graham, Devon Quick, and Jon Runyeon. "13.1 Sensory Receptors." *Anatomy & Physiology*. 2020. Open Oregon State, Oregon State University. https://open.oregonstate.education/aandp /chapter/13-1-sensory-receptors/.

BodModz. "Cute Girl Having a TONGUE SPLIT (Full Procedure)." Video. YouTube, December 13, 2014. https://youtu.be/-cMG2QyN-mE.

Boggs, J. M. "Myelin Basic Protein: A Multifunctional Protein." *Cellular and Molecular Life Sciences* 63, no. 17 (2006): 1945–1961. doi:10.1007/s00018-006-6094-7.

Corder, Gregory, Biafra Ahanonu, Benjamin F. Grewe, Dong Wang, Mark J. Schnitzer, and Grégory Scherrer. "An Amygdalar Neural Ensemble That Encodes the Unpleasantness of Pain." *Science* 363, no. 6424 (2019): 276–281. doi:10.1126/science.aap8586.

Dubin, Adrienne E., and Ardem Patapoutian. "Nociceptors: The Sensors of the Pain Pathway." *Journal of Clinical Investigation* 120, no. 11 (2010): 3760–3772. doi:10.1172/jci42843.

Foell, Jens. "Phantom Limbs and Perceived Pain." TED Talk, 12:22, TEDxFSU, published September 22, 2017. https://www.youtube.com /watch?v=7-EpZcKgufM.

Foell, Jens, Robin Bekrater-Bodmann, Herta Flor, and Jonathan Cole. "Phantom Limb Pain After Lower Limb Trauma." *International Journal of Lower Extremity Wounds* 10, no. 4 (2011): 224–235. doi:10.1177/1534734611428730.

Heckert, Justin. 2012. "The Hazards of Growing Up Painlessly." *New York Times Magazine*, November 18, 2012. https://www.nytimes.com/2012/11/18/magazine/ashlyn-blocker-feels-no-pain.html.

Jabr, Ferris. "Know Your Neurons: How to Classify Different Types of Neurons in the Brain's Forest." Brainwaves (blog), *Scientific American*, May 16, 2012. https://blogs.scientificamerican.com/brainwaves/know-your-neurons-classifying-the-many-types-of-cells-in-the-neuron-forest/.

Kamping, Sandra, Jamila Andoh, Isabelle C. Bomba, Martin Diers, Eugen Diesch, and Herta Flor. "Contextual Modulation of Pain in Masochists." *PAIN* 157, no. 2 (2016): 445–455. doi:10.1097/j.pain.0000000000000390.

Kwon, Diana. "The Battle over Pain in the Brain." *Scientific American*, April 28, 2016. https://www.scientificamerican.com/article/the-battle-over-pain-in-the-brain/.

Leknes, Siri, and Irene Tracey. "A Common Neurobiology for Pain and Pleasure." *Nature Reviews Neuroscience* 9, no. 4 (2008): 314–320. doi:10.1038/nrn2333.

Marzvanyan, Anna, and Ali Alhawaj. "Physiology, Sensory Receptors." In *StatPearls*. Treasure Island, FL: StatPearls Publishing, 2020. https://www.ncbi.nlm.nih.gov/books/NBK539861/.

Mayer, Emeran A., and M. Catherine Bushnell. *Functional Pain Syndromes: Presentation and Pathophysiology*. Seattle, WA: IASP Press, 2009.

Nagasako, Elna M., Anne Louise Oaklander, and Robert H. Dworkin. "Congenital Insensitivity to Pain: An Update." *Pain* 101, no. 3 (2003): 213–219. doi:10.1016/s0304-3959(02)00482-7.

Poldrack, Russell A., Jeanette A. Mumford, and Thomas E. Nichols. *Handbook of Functional MRI Data Analysis*. New York: Cambridge University Press, 2012.

Riedel, W., and G. Neeck. "Nociception, Pain, and Antinociception: Current Concepts." *Zeitschrift Fur Rheumatologie* 60, no. 6 (2001): 404–415. doi:10.1007/s003930170003.

Wall, Patrick. *PAIN: The Science of Suffering*. New York: Columbia University Press, 2000.

Wark, Barry, Brian Nils Lundstrom, and Adrienne Fairhall. "Sensory Adaptation." *Current Opinion in Neurobiology* 17, no. 4 (2007): 423–429. doi:10.1016/j.conb.2007.07.001.

Whittington, Camilla, and Katherine Belov. "Platypus Venom: A Review." *Australian Mammalogy* 29, no. 1 (2007): 57. doi:10.1071/am07006.

The Fuffering of the Faints

Barbarich-Marsteller, Nicole C., Richard W. Foltin, and B. Timothy Walsh. "Does Anorexia Nervosa Resemble an Addiction?" *Current Drug Abuse Reviews* 4, no. 3 (2011): 197–200.

Bastian, Brock. "The Masochism Tango." *The Economist*, February 3, 2011. https://www.economist.com/science-and-technology/2011/02/03/the-masochism-tango.

Bastian, Brock. *The Other Side of Happiness*. Bungay, UK: Penguin Random House UK, 2018.

Bastian, Brock, Jolanda Jetten, and Fabio Fasoli. "Cleansing the Soul by Hurting the Flesh." *Psychological Science* 22, no. 3 (2011): 334–335. doi:10.1177/0956797610397058.

Bataille, Georges, and Peter Connor. *The Tears of Eros*. San Francisco: City Lights Books, 1989. First published 1961 by Jacques Pauvert, Paris.

Baumeister, Roy F. *Escaping the Self: Alcoholism, Spirituality, Masochism, and Other Flights from the Burden of Selfhood*. New York: Basic Books, 1991.

Chen, Andrew H. *Flagellant Confraternities and Italian Art, 1260–1610*. Amsterdam: Amsterdam University Press, 2018.

Clifford, Catherine. "Billionaire Jack Dorsey's 11 'Wellness' Habits: From No Food All Weekend to Ice Baths." CNBC, April 8, 2019. https://www.cnbc.com/2019/04/08"/twitter-and-square-ceo-jack-dorsey-on-his-personal-wellness-habits.html.

Cohn, Samuel K., Jr. "The Black Death: End of a Paradigm." *American Historical Review* 107, no. 3 (2002). doi:10.1086/ahr/107.3.703.

Cooper, William M. *Flagellation & the Flagellants: A History of the Rod*. Amsterdam: Fredonia Books, 2001.

DeBoer, Scott, Allen Falkner, Troy Amundson, Myrna Armstrong, Michael Seaver, Steve Joyner, and Lisa Rapoport. "Just Hanging Around: Questions and Answers About Body Suspensions." *Journal of Emergency Nursing* 34, no. 6 (2008): 523–529. doi:10.1016/j.jen.2007.10.014.

"Eating Disorder Statistics: General & Diversity Stats." National Association of Anorexia Nervosa and Associated Disorders. https://anad.org/get-informed/about-eating-disorders/eating-disorders-statistics/.

The Exhibition of Female Flagellants: Parts One and Two (English Flagellant Fiction 1770–1830). Birchgrove Press, 2012. First published 1777, G. Peacock, London.

Forsyth, Craig J., and Jessica Simpson. "Everything Changes Once You Hang: Flesh Hook Suspension." *Deviant Behavior* 29, no. 4 (2008): 367–387. doi:10.1080/01639620701588305.

Glucklich, Ariel. *Sacred Pain: Hurting the Body for the Sake of the Soul*. New York: Oxford University Press, 2001.

Gobel, Eric A. "Liturgical Processions of the Black Death." *Hilltop Review* 9, no. 2 (2017): 32–46.

Gregg, John R. *Sex, the Illustrated History: Through Time, Religion and Culture: Volume III*. Bloomington, IN: Xlibris Corporation, 2016

Hope, Valerie M. *Roman Death*. London: Bloomsbury Publishing, 2009.

Jean, A., G. Conductier, C. Manrique, C. Bouras, P. Berta, R. Hen, Y. Charnay, J. Bockaert, and V. Compan. "Anorexia Induced by Activation of Serotonin 5-HT4

Receptors Is Mediated by Increases in CART in the Nucleus Accumbens." *Proceedings of the National Academy of Sciences* 104, no. 41 (2007): 16335–16340. doi:10.1073/pnas.0701471104.

Konnikova, Maria. "Pain Really Does Make Us Gain." *New Yorker*, December 24, 2014. https://www.newyorker.com/science/maria-konnikova/pain-really-make-us-gain.

Lolme, Jean Louis de, Jacques Boileau, and Henry Layard. *The History of the Flagellants: Otherwise, of Religious Flagellations Among Different Nations, and Especially Among Christians. Being a Paraphrase and Commentary on the Historia Flagellantium of the Abbé Boileau, Doctor of the Sorbonne, Canon of the Holy Chapel, & C.* London: Printed for G. Robinson, No. 25, Paternoster Row, 1783.

Marshall, Wyatt. "The Therapeutic Experience of Being Suspended by Your Skin." *The Atlantic*, September 21, 2012. https://www.theatlantic.com/health/archive/2012/09/the-therapeutic-experience-of-being-suspended-by-your-skin/262644/.

Martin, Seán. *The Black Death*. Edison, NJ: Chartwell Books, 2007.

Morris, David B. *The Culture of Pain*. Berkeley: University of California Press, 2007.

Murphy, Sean C., and Brock Bastian. "Emotionally Extreme Life Experiences Are More Meaningful." *Journal of Positive Psychology* 15, no. 4 (2019): 531–542. doi:10.1080/17439760.2019.1639795.

"The Pilgrimage of the Flagellants About Georg Doring's Geiselfahrt," *Blackwood's Edinburgh Magazine*, 1833.

Sandburg, Carl. "Fog." In *Chicago Poems*. New York: Henry Holt, 1916.

Sanger, Gareth J., Per M. Hellström, and Erik Näslund. "The Hungry Stomach: Physiology, Disease, and Drug Development Opportunities." *Frontiers in Pharmacology* 1 (2011). doi:10.3389/fphar.2010.00145.

Tischauser, Leslie V. "Okipa." *Encyclopedia of the Great Plains*, 2011. http://plainshumanities.unl.edu/encyclopedia/doc/egp.rel.037.

Tuchman, Barbara W. *A Distant Mirror*. New York: Random House, 1978.

Whitehouse, Harvey, and Jonathan A. Lanman. "The Ties That Bind Us." *Current Anthropology* 55, no. 6 (2014): 674–695. doi:10.1086/678698.

Lava Mouth

Bastian, Brock. *The Other Side of Happiness*. Bungay, UK: Penguin Random House UK, 2018.

Byrnes, Nadia K., and John E. Hayes. "Behavioral Measures of Risk Tasking, Sensation Seeking and Sensitivity to Reward May Reflect Different Motivations for Spicy Food Liking and Consumption." *Appetite* 103 (2016): 411–422. doi:10.1016/j.appet.2016.04.037.

———. "Personality Factors Predict Spicy Food Liking and Intake." *Food Quality and Preference* 28, no. 1 (2013): 213–221. doi:10.1016/j.foodqual.2012.09 .008.

Caterina, Michael J., Mark A. Schumacher, Makoto Tominaga, Tobias A. Rosen, Jon D. Levine, and David Julius. "The Capsaicin Receptor: A Heat-Activated Ion Channel in the Pain Pathway." *Nature* 389, no. 6653 (1997): 816–824. doi:10.1038/39807.

Ettenberg, Jodi. "A Brief History of Chili Peppers." Legal Nomads, May 17, 2020. https://www.legalnomads.com/history-chili-peppers/.

Fett, Debra D. "Botanical Briefs: Capsicum Peppers." *Cutis* 72, no. 1 (2003): 21–23.

"Final Report on the Safety Assessment of Capsicum Annuum Extract, Capsicum Annuum Fruit Extract, Capsicum Annuum Resin, Capsicum Annuum Fruit Powder, Capsicum Frutescens Fruit, Capsicum Frutescens Fruit Extract, Capsicum Frutescens Resin, and Capsaicin." *International Journal of Toxicology* 26 (Suppl. 1, 2007): 3–106. doi:10.1080/10915810601163939.

Fitzgerald, Maria. "Capsaicin and Sensory Neurones—a Review." *Pain* 15, no. 1 (1983): 109–130. doi:10.1016/0304-3959(83)90012-x.

Kimball, Tracy. "Fire in the Belly: Pepper-Eating Contest in Fort Mill Was Almost a Dead Heat." *The Herald* (Rock Hill, SC), August 14, 2019. https://www .heraldonline.com/news/local/article233504922.html.

Lee, Jin-Seong, Sung-Gon Kim, Hyeun-Kyeung Kim, Sun-Yong Baek, and Cheol-Min Kim. "Acute Effects of Capsaicin on Proopioimelanocortin mRNA Levels in the Arcuate Nucleus of Sprague-Dawley Rats." *Psychiatry Investigation* 9, no. 2 (2012): 187. doi:10.4306/pi.2012.9.2.187.

Meyer, Zlati. "Hot Sauce Industry Sets Tongues—and Sales—Ablaze." *USA Today*, July 30, 2017. https://www.usatoday.com/story/money/2017/07/30 /hot-sauce-industry-fire-supermarkets-mcdonalds/519660001/.

Nmaju, Anyauba Uduka, Iwasam Ekom Joshua, Udemeobong Edet Okon, Azubuike Amakwe Nwankwo, and Eme Efiom Osim. "Long-Term Consumption of Capsicum Annum (Chilli Pepper) and Capsaicin Diets Suppresses Pain Perception and Improves Social Behaviour of CD-1 Mice." *Nutrition & Food Science* 48, no. 6 (2018): 911–921. doi:10.1108/nfs-02-2018 -0054.

Nolden, Alissa A., and John E. Hayes. "Perceptual and Affective Responses to Sampled Capsaicin Differ by Reported Intake." *Food Quality and Preference* 55 (2017): 26–34. doi:10.1016/j.foodqual.2016.08.003.

Park, Thomas J., Ying Lu, René Jüttner, Ewan St. J. Smith, Jing Hu, Antje Brand, Christiane Wetzel, et al. "Selective Inflammatory Pain Insensitivity in the African Naked Mole-Rat (Heterocephalus Glaber)." *PloS Biology* 6, no. 1 (2008): e13. doi:10.1371/journal.pbio.0060013.

Price, R. C., W. Gandhi, C. Nadeau, R. Tarnavskiy, A. Qu, E. Fahey, L. Stone, and P. Schweinhardt. "Characterization of a Novel Capsaicin/Heat Ongoing Pain Model." *European Journal of Pain* 22, no. 2 (2017): 370–384. doi:10.1002 /ejp.1126.

Rozin, Paul, Lily Guillot, Katrina Fincher, Alexander Rozin, and Eli Tsukayama. "Glad to Be Sad, and Other Examples of Benign Masochism." *Judgement and Decision Making* 8, no. 4 (2013): 439–447.

Yang, Fan, and Jie Zheng. "Understand Spiciness: Mechanism of TRPV1 Channel Activation by Capsaicin." *Protein & Cell* 8, no. 3 (2017): 169–177. doi:10.1007 /s13238-016-0353-7.

The Name of Things

American Psychiatric Association. *Diagnostic and Statistical Manual of Mental Disorders*. 5th ed. Arlington, VA: American Psychiatric Association, 2013.

Barry, Dave. "Dave Barry Learns Everything You Need to Know About Being a Husband From Reading *50 Shades of Grey*." *Time*, March 4, 2014. https://time .com/3030375/dave-barry-50-shades-of-grey/.

The Birchen Bouquet. London: Locus Elm Press, 2017. First published around 1770.

Braun, Adee. "Looking to Quell Sexual Urges? Consider the Graham Cracker." *The Atlantic*, January 15, 2014. https://www.theatlantic.com/health/archive/2014/01 /looking-to-quell-sexual-urges-consider-the-graham-cracker/282769/.

Cleugh, James. *The First Masochist*. New York: Stein and Day, 1967.

Conliffe, Ciaran. "Leopold Von Sacher-Masoch, Poet of Masochism." HeadStuff, June 8, 2015. https://www.headstuff.org/culture/literature/leopold-von-sacher -masoch-poet-of-masochism/.

Cuccinello, Hayley. "Fifty Shades of Green: How Fanfiction Went from Dirty Little Secret to Money Machine." *Forbes*, February 10, 2017. https://www.forbes .com/sites/hayleycuccinello/2017/02/10/fifty-shades-of-green-how-fanfiction -went-from-dirty-little-secret-to-money-machine/?sh=7c22badf264c.

Deleuze, Gilles. *Coldness and Cruelty*. New York: Zone Books, 1967.

Eakin, Emily. "Grey Area: How 'Fifty Shades' Dominated the Market." *New York Review of Books*, July 27, 2012. https://www.nybooks.com/daily/2012/07/27 /seduction-and-betrayal-twilight-fifty-shades/.

Englishwoman's Domestic Magazine. Supplement on Domestic Disciplinary Actions, 1870.

An Expert. *The Romance of Chastisement; Or, Revelations of the School and Bedroom*. Birchgrove Press, 2011. First published 1886.

Falaky, Fayçal. *Social Construct, Masochist Contract: Aesthetics of Freedom and Submission in Rousseau*. Albany: State University of New York Press, 2014.

"Fifty Shades Darker (2017)." Rotten Tomatoes. https://www.rottentomatoes .com/m/fifty_shades_darker.

"Fifty Shades Freed (2018)." Rotten Tomatoes. https://www.rottentomatoes
.com/m/fifty_shades_freed.

"Fifty Shades of Grey (2015)." Rotten Tomatoes. https://www.rottentomatoes
.com/m/fifty_shades_of_grey.

Frances, Allen. *Saving Normal: An Insider's Revolt Against Out-of-Control Psychiatric Diagnosis, DSM-5, Big Pharma, and the Medicalization of Ordinary Life.* New York: William Morrow, 2013.

Freud, Sigmund, and James Strachey. *Three Essays on the Theory of Sexuality.* Welwyn Garden City, UK: Alcuin Press, 1949.

Freud, Sigmund, James Strachey, Gregory Zilboorg, and Peter Gay. *Beyond the Pleasure Principle.* New York: W. W. Norton & Company, 1961.

James, E. L. *Fifty Shades of Grey.* New York: Random House, 2011.

Jones, James H. *Alfred C. Kinsey.* New York: W. W. Norton & Company, 1997.

Krafft-Ebing, Richard von, and Franklin S. Klaf. *Psychopathia Sexualis.* New York: Arcade Publishing, 1886.

Lutz, Deborah. *Pleasure Bound: Victorian Sex Rebels and the New Eroticism.* New York: W. W. Norton & Company, 2011.

McLynn, Kim. "'Fifty Shades of Grey' Was the Best-Selling Book of the Decade in the U.S." NPD Group, December 18, 2019. https://www.npd.com/wps/portal /npd/us/news/press-releases/2019/fifty-shades-of-grey-was-the-best-selling-book -of-the-decade-in-the-us-the-npd-group-says/.

Oosterhuis, Harry. *Stepchildren of Nature: Krafft-Ebing, Psychiatry, and the Making of Sexual Identity.* Chicago: University of Chicago Press, 2000.

Palmore, Erdman. "Published Reactions to the Kinsey Report." *Social Forces* 31, no. 2 (1952): 165–172.

Raye, Martha. "Ooh, Doctor Kinsey!" *Ooh, Dr. Kinsey! / After You've Gone.* Produced by Phil Moore. 1949.

Rousseau, Jean-Jacques. *The Confessions of Jean-Jacques Rousseau.* New Delhi, India: Gopsons Papers, 2000. First published 1782.

Sacher-Masoch, Leopold von. *Venus in Furs.* New York: Zone Books, 1870.

Sacher-Masoch, Wanda von. *The Confessions of Wanda von Sacher-Masoch.* San Francisco: Re/Search Publications, 1990. First published 1906.

Whyte, Marama. "I Read 'Fifty Shades of Grey' and It Was Worse Than I Imagined." Hypable, February 12, 2015. https://www.hypable.com /fifty-shades-of-grey-book-review/.

Williams, Zoe. "Why Women Love Fifty Shades of Grey." *The Guardian*, July 6, 2012. https://www.theguardian.com/books/2012/jul/06 /why-women-love-fifty-shades-grey.

When the Lights Go Dark

American Psychiatric Association. *Diagnostic and Statistical Manual of Mental Disorders.* 5th ed. Arlington, VA: American Psychiatric Association, 2013.

Cleare, Seonaid, Andrew Gumley, and Rory C. O'Connor. "Self-Compassion, Self-Forgiveness, Suicidal Ideation, and Self-Harm: A Systematic Review." *Clinical Psychology & Psychotherapy* 26, no. 5 (2019): 511–530. doi:10.1002/cpp .2372.

"Eating Disorder Statistics: General & Diversity Stats." National Association of Anorexia Nervosa and Associated Disorders. https://anad.org/get-informed /about-eating-disorders/eating-disorders-statistics/.

Edmondson, Amanda J., Cathy A. Brennan, and Allan O. House. "Non-Suicidal Reasons for Self-Harm: A Systematic Review of Self-Reported Accounts." *Journal of Affective Disorders* 191 (2016): 109–117. doi:10.1016/j.jad.2015.11 .043.

Gratz, Kim L. "Measurement of Deliberate Self-Harm: Preliminary Data on the Deliberate Self-Harm Inventory." *Journal of Psychopathy and Behavioral Assessment* 23, no. 4 (2001): 253–263.

Skegg, Keren. "Self-Harm." *The Lancet* 366 (2005): 1471–1483.

van der Kolk, Bessel. *The Body Keeps the Score*. United States: Penguin, 2015.

Social Creatures

Baimel, Adam, Susan A. J. Birch, and Ara Norenzayan. "Coordinating Bodies and Minds: Behavioral Synchrony Fosters Mentalizing." *Journal of Experimental Social Psychology* 74 (2018): 281–290. doi:10.1016/j.jesp.2017.10.008.

Barber, Nigel. "Taking One's Cue from Others." The Human Beast (blog), *Psychology Today*, May 31, 2017. https://www.psychologytoday.com/us/blog /the-human-beast/201705/taking-ones-cue-others.

Boecker, H., T. Sprenger, M. E. Spilker, G. Henriksen, M. Koppenhoefer, K. J. Wagner, M. Valet, A. Berthele, and T. R. Tolle. "The Runner's High: Opioidergic Mechanisms in the Human Brain." *Cerebral Cortex* 18, no. 11 (2008): 2523–2531. doi:10.1093/cercor/bhn013.

Carney, Scott. *What Doesn't Kill Us*. New York: Rodale Books, 2017.

Christie, Macdonald J., and Gregory B. Chesher. "Physical Dependence on Physiologically Released Endogenous Opiates." *Life Sciences* 30, no. 14 (1982): 1173–1177.

Cohen, Emma E. A., Robin Ejsmond-Frey, Nicola Knight, and R. I. M. Dunbar. "Rowers' High: Behavioural Synchrony Is Correlated with Elevated Pain Thresholds." *Biology Letters* 6, no. 1 (2009): 106–108. doi:10.1098/rsbl.2009 .0670.

Cuadros, Zamara, Esteban Hurtado, and Carlos Cornejo. "Measuring Dynamics of Infant-Adult Synchrony Through Mocap." *Frontiers in Psychology* 10 (2019). doi:10.3389/fpsyg.2019.02839.

Darley, John M., and Bibb Latane. "Bystander Intervention in Emergencies: Diffusion of Responsibility." Pt. 1. *Journal of Personality and Social Psychology* 8, no. 4 (1968): 377–383. doi:10.1037/h0025589.

Davidson, Nick. "Cold Plunge." *Outside*, May 10, 2011. https://www.outsideonline.com /1871876/cold-plunge.

Feldman, Ruth. "Parent–Infant Synchrony." *Current Directions in Psychological Science* 16, no. 6 (2007): 340–345. doi:10.1111/j.1467-8721.2007.00532.x.

"4 Army Ranger Candidates Die in Chilly Florida Swamp." *New York Times*, February 17, 1995. https://www.nytimes.com/1995/02/17/us/4-army-ranger -candidates-die-in-chilly-florida-swamp.html.

Fuss, Johannes, Jörg Steinle, Laura Bindila, Matthias K. Auer, Hartmut Kirchherr, Beat Lutz, and Peter Gass. "A Runner's High Depends on Cannabinoid Receptors in Mice." *Proceedings of the National Academy of Sciences* 112, no. 42 (2015): 13105–13108. doi:10.1073/pnas.1514996112.

Galbusera, Laura, Michael T. M. Finn, Wolfgang Tschacher, and Miriam Kyselo. "Interpersonal Synchrony Feels Good but Impedes Self-Regulation of Affect." *Scientific Reports* 9, no. 1 (2019). doi:10.1038/s41598-019-50960-0.

Guindon, Josee, and Andrea Hohmann. "The Endocannabinoid System and Pain." *CNS & Neurological Disorders—Drug Targets* 8, no. 6 (2009): 403–421. doi:10.2174/187152709789824660.

Leclère, Chloë, Sylvie Viaux, Marie Avril, Catherine Achard, Mohamed Chetouani, Sylvain Missonnier, and David Cohen. "Why Synchrony Matters During Mother-Child Interactions: A Systematic Review." *PloS ONE* 9, no. 12 (2014): e113571. doi:10.1371/journal.pone.0113571.

Lewis, Zachary, and Philip J. Sullivan. "The Effect of Group Size and Synchrony on Pain Threshold Changes." *Small Group Research* 49, no. 6 (2018): 723–738. doi:10.1177/1046496418765678.

Mead, Rebecca. "The Subversive Joy of Cold-Water Swimming." *New Yorker*, January 27, 2020. https://www.newyorker.com/magazine/2020/01/27/the -subversive-joy-of-cold-water-swimming.

Miller, Greg. "How Movies Synchronize the Brains of an Audience." *Wired*, August 28, 2014. https://www.wired.com/2014/08/cinema-science-mind-meld/.

Moore, Melissa. "How the Endocannabinoid System Was Discovered." Labroots, April 5, 2018. https://www.labroots.com/trending/cannabis-sciences/8456 /endocannabinoid-system-discovered.

Richardson, Jennelle Durnett. "Cannabinoids Modulate Pain by Multiple Mechanisms of Action." *Journal of Pain* 1, no. 1 (2000): 2–14. doi:10.1016 /s1526-5900(00)90082-8.

Scarry, Elaine. *The Body in Pain: The Making and Unmaking of the World*. New York: Oxford University Press, 1985.

"A Science Odyssey: People and Discoveries: Role of Endorphins Discovered, 1975." PBS. https://www.pbs.org/wgbh/aso/databank/entries/dh75en .html.

Sharon-David, Hilla, Moran Mizrahi, Michal Rinott, Yulia Golland, and Gurit E. Birnbaum. "Being on the Same Wavelength: Behavioral Synchrony

Between Partners and Its Influence on the Experience of Intimacy." *Journal of Social and Personal Relationships* 36, no. 10 (2018): 2983–3008. doi:10.1177/0265407518809478.

Sprouse-Blum, Adam S., Greg Smith, Daniel Sugai, and F. Don Parsa. "Understanding Endorphins and Their Role in Pain Management." *Hawai'i Medical Journal* 69, no. 3 (2010): 70–71. https://www.ncbi.nlm.nih.gov/pmc/articles /PMC3104618/.

Stefano, George B., Yannick Goumon, Federico Casares, Patrick Cadet, Gregory L. Fricchione, Christos Rialas, Doris Peter, et al. "Endogenous Morphine." *Trends in Neurosciences* 23, no. 9 (2000): 436–442. doi:10.1016 /s0166-2236(00)01611-8.

Tipton, M. J., N. Collier, H. Massey, J. Corbett, and M. Harper. "Cold Water Immersion: Kill or Cure?" *Experimental Physiology* 102, no. 11 (2017): 1335–1355. doi:10.1113/ep086283.

Weeks, Jonny. "A Cold-Water Cure? My Weekend with the 'Ice Man.'" *The Guardian*, May 8, 2019. https://www.theguardian.com/world/2019/may/08 /wim-hof-cold-water-immersion-cure-ice-man-outdoor-swimming.

The Ultramarathon

Barker, Sarah. "Ultrarunner Courtney Dauwalter Takes on the World's Most Sadistic Endurance Race." Deadspin, November 9, 2018. https://dead spin.com/ultrarunner-courtney-dauwalter-takes-on-the-worlds-most-18301 36537.

Bramble, Dennis M., and Daniel E. Lieberman. "Endurance Running and the Evolution of *Homo*." *Nature* 432, no. 7015 (2004): 345–352. doi:10.1038 /nature03052.

Carroll, Larry. "2019 Big Dog Backyard Preview: 'This Is a Race to the Death.'" Irun4ultra. https://irun4ultra.com/2019-big-dog-backyard-preview -this-is-a-race-to-the-death/.

Cohen, Daniel C., Alison Winstanley, Alec Engledow, Alastair C. Windsor, and James R. Skipworth. "Marathon-Induced Ischemic Colitis: Why Running Is Not Always Good for You." *American Journal of Emergency Medicine* 27, no. 2 (2009): 255.e5–255.e7. doi:10.1016/j.ajem.2008.06.033.

Cowart, Leigh. "Ultra Pain: The Insane Things That People Put Themselves Through to Complete 100-Mile Marathons." SBNation, July 10, 2014. https://www.sbnation.com/2014/7/10/5887187/ultra-pain-the-insane-things -that-insane-people-put-themselves.

———. "Why Running Sometimes Makes You Shit Blood." Deadspin, September 1, 2017. https://deadspin.com/why-running-sometimes-makes -you-shit-blood-1798536358.

Crockett, Davy. "Man vs. Horse—Racing Ultradistances." Ultrarunning History, July 30, 2018. http://ultrarunninghistory.com/man-vs-horse/.

———. "Yiannis Kouros—Greek Greatness." Ultrarunning History, June 9, 2019. https://ultrarunninghistory.com/yiannis-kouros/.

Fallon, K. E., G. Sivyer, K. Sivyer, and A. Dare. "The Biochemistry of Runners in a 1600 km Ultramarathon." *British Journal of Sports Medicine* 33, no. 4 (1999): 264–269. doi:10.1136/bjsm.33.4.264.

Faress, Ahmed. "'Runs' from a Run: A Case of Exercise Induced Ischemic Colitis." *World Journal of Emergency Medicine* 8, no. 4 (2017): 302. doi:10.5847 /wjem.j.1920-8642.2017.04.010.

Finn, Adharanand. "When 26.2 Miles Just Isn't Enough—the Phenomenal Rise of the Ultramarathon." *The Guardian*, April 2, 2018. https://www.the guardian.com/lifeandstyle/2018/apr/02/ultrarunner-ultramarathon-racing-100 -miles.

Fox, Kit. "Meet the Man Who Ran Almost 250 Miles with a Smile on His Face." *Men's Journal*, 2017. https://www.mensjournal.com/sports /meet-the-winner-of-the-worlds-cruelest-ultramarathon-w510640/.

Geisler, Maria, Luise Eichelkraut, Wolfgang H. R. Miltner, and Thomas Weiss. "An fMRI Study on Runner's High and Exercise-Induced Hypoalgesia After a 2-h Run in Trained Non-Elite Male Athletes." *Sport Sciences for Health* 16, no. 1 (2019): 159–167. doi:10.1007/s11332-019-00592-8.

Given, Karen. "A Horse Race Without a Horse: How Modern Trail Ultramarathoning Was Invented." WBUR, June 28, 2019. https://www.wbur.org /onlyagame/2019/06/28/ultramarathon-gordon-ainsleigh-western-states.

Huber, Martin Fritz. "The Existential Torture of a Race with No End." *Outside*, October 26, 2018. https://www.outsideonline.com/2358936 /bigs-backyard-ultra-existential-torture.

Iltis, Annika, and Timothy James Kane. *The Barkley Marathons: The Race That Eats Its Young.* DVD. 2014.

Lake, Lazarus. "Hourly Public Posts During the 2019 Big's Backyard Ultra." Facebook, 2019. http://facebook.com/lazarus.lake.

Longman, Jer. "The Marathon's Random Route to Its Length." *New York Times*, April 21, 2012. https://www.nytimes.com/2012/04/21/sports/the-marathons -accidental-route-to-26-miles-385-yards.html.

Pearson, Andy. "Big's Backyard Ultra Gets Bigger and Bigger." *Trail Runner*, October 31, 2018. https://trailrunnermag.com/races/bigs-backyard-ultra-gets-bigger -and-bigger.html.

Proctor, Dave. "Race Report—Big's Backyard Ultra." Outrun Rare (blog), October 24, 2019. https://outrunrare.com/race-report-bigs-backyard-ultra/.

Raichlen, D. A., A. D. Foster, G. L. Gerdeman, A. Seillier, and A. Giuffrida. "Wired to Run: Exercise-Induced Endocannabinoid Signaling in Humans and Cursorial Mammals with Implications for the 'Runner's High.'" *Journal of Experimental Biology* 215, no. 8 (2012): 1331–1336. doi:10.1242/jeb .063677.

Schulkin, Jay. "Evolutionary Basis of Human Running and Its Impact on Neural Function." *Frontiers in Systems Neuroscience* 10, no. 59 (2016). doi:10.3389
/fnsys.2016.00059.

Simpson, Duncan, Phillip G. Post, Greg Young, and Peter R. Jensen. "'It's Not
About Taking the Easy Road': The Experiences of Ultramarathon Runners."
Sport Psychologist 28, no. 2 (2014): 176–185. doi:10.1123/tsp.2013-0064.

"35th Edition, October 01 to 11, 2021, 250 km in 7 Days." Legendary Marathon
des Sables. https://www.marathondessables.com/en.

University of Utah. "How Running Made Us Human: Endurance Running Let Us
Evolve to Look the Way We Do." Science Daily, November 24, 2004. https://
www.sciencedaily.com/releases/2004/11/041123163757.htm.

Vigneron, Peter. "A Pheidippides F.A.Q." *Runner's World*, September 13, 2010. https://
www.runnersworld.com/runners-stories/a20787158/a-pheidippides-f-a-q/.

Serious Playtime

Burch, Rebecca L., and Catherine Salmon. "The Rough Stuff: Understanding
Aggressive Consensual Sex." *Evolutionary Psychological Science* 5, no. 4 (2019):
383–393. doi:10.1007/s40806-019-00196-y.

Dunkley, Cara R., Craig D. Henshaw, Saira K. Henshaw, and Lori A. Brotto.
"Physical Pain as Pleasure: A Theoretical Perspective." *Journal of Sex Research* 57,
no. 4 (2019): 421–437. doi:10.1080/00224499.2019.1605328.

Easton, Dossie, and Janet W. Hardy. *The New Bottoming Book*. Emeryville, CA:
Greenery Press, 2001.

———. *The New Topping Book*. Oakland, CA: Greenery Press, 2003.

History of Sexual Punishment in Pictures. Frankfurt, Germany: Goliath, 2019.

"IASP Announces Revised Definition of Pain." International Association for the
Study of Pain, July 16, 2020. https://www.iasp-pain.org/PublicationsNews
/NewsDetailaspx?ItemNumber=10475.

Kroll, Eric. *John Willie's Best of Bizarre*. Italy: Taschen, 2001.

Musser, Amber Jamilla. *Sensational Flesh: Race, Power, and Masochism*. New York:
New York University Press, 2014.

Newmahr, Staci. *Playing on the Edge: Sadomasochism, Risk, and Intimacy*. Bloomington: Indiana University Press, 2011.

———. "Rethinking Kink: Sadomasochism as Serious Leisure." *Qualitative Sociol-
ogy* 33, no. 3 (2010): 313–331. doi:10.1007/s11133-010-9158-9.

Phillips, Anita. *In Defense of Masochism*. New York: St. Martin's Press, 1998.

Weierstall, Roland, and Gilda Giebel. "The Sadomasochism Checklist: A Tool for
the Assessment of Sadomasochistic Behavior." *Archives of Sexual Behavior* 46,
no. 3 (2016): 735–745. doi:10.1007/s10508-016-0789-0.

Williams, Amanda C. de C., and Kenneth D. Craig. "Updating the Definition of Pain." *PAIN* 157, no. 11 (2016): 2420–2423. doi:10.1097/j
.pain.0000000000000613.

INDEX

Katrina Noell, 10th Moon Photography

Leigh Cowart is a researcher and journalist whose work has appeared in the *Washington Post*, *New York Magazine*, BuzzFeed News, Hazlitt, Longreads, Vice, and other outlets. Before becoming a journalist, Cowart was immersed in academia, doing research on subjects like sexual dimorphism in leaf-nosed bats and resource allocation in flowers. They live in Asheville, North Carolina, with their family and their feline office manager, Larry Hotdogs.

PublicAffairs is a publishing house founded in 1997. It is a tribute to the standards, values, and flair of three persons who have served as mentors to countless reporters, writers, editors, and book people of all kinds, including me.

I. F. STONE, proprietor of *I. F. Stone's Weekly*, combined a commitment to the First Amendment with entrepreneurial zeal and reporting skill and became one of the great independent journalists in American history. At the age of eighty, Izzy published *The Trial of Socrates*, which was a national bestseller. He wrote the book after he taught himself ancient Greek.

BENJAMIN C. BRADLEE was for nearly thirty years the charismatic editorial leader of *The Washington Post*. It was Ben who gave the *Post* the range and courage to pursue such historic issues as Watergate. He supported his reporters with a tenacity that made them fearless and it is no accident that so many became authors of influential, best-selling books.

ROBERT L. BERNSTEIN, the chief executive of Random House for more than a quarter century, guided one of the nation's premier publishing houses. Bob was personally responsible for many books of political dissent and argument that challenged tyranny around the globe. He is also the founder and longtime chair of Human Rights Watch, one of the most respected human rights organizations in the world.

·　　·　　·

For fifty years, the banner of Public Affairs Press was carried by its owner Morris B. Schnapper, who published Gandhi, Nasser, Toynbee, Truman, and about 1,500 other authors. In 1983, Schnapper was described by *The Washington Post* as "a redoubtable gadfly." His legacy will endure in the books to come.

Peter Osnos, *Founder*